'Succeeding in business is about managing risk, controlling all the factors that are controllable and constantly improving your competitive advantage. At Open Country Dairy we live by this mantra and Lean Management is integral to everything we do. I cannot commend this book enough to all farmers and business people that strive to improve and grow successful enterprises. An outstanding contribution to NZ farming.'

—Laurie Margrain, Chairman, Open Country Dairy Ltd

'Within GDF we aim to inspire leading farmers. From a personal experience, I am convinced that the Lean approach is very helpful to improve management. So if you want to improve as a dairy farmer, read this book full of practical tools and experiences.'

—Ad van Velde, President, Global Dairy Farmers,
CEO, DairyNext, Owner, Hunsingo Dairy

T0315492

the Lean Dairy Farm

the
Lean Dairy
Farm

Eliminate Waste, Save Time, Cut Costs

**Creating a More Productive,
Profitable and
Higher Quality Farm**

JANA HOCKEN WITH **MAT HOCKEN**

WILEY

First published in 2019 by John Wiley & Sons Australia, Ltd
42 McDougall St, Milton Qld 4064

Office also in Melbourne

Typeset in 11/15pt Berkeley Oldstyle

© John Wiley & Sons Australia, Ltd 2019

The moral rights of the author have been asserted

A catalogue record for this
book is available from the
National Library of Australia

Cover design by Wiley

Cover image (upper): © Photos by R A Kearton / Getty Images
Cover image (lower): Javier Vargas

10 9 8 7 6 5 4 3 2 1

Contents

PART III
The culture of a Lean Farm 371

Acknowledgements

I would like to thank the whole team at Grassmere Dairy for their contribution, patience and support in completing this book (one of whom took the photo on the cover).

Thank you to Lucy, Chris and the team at Wiley for turning this book into a reality. My editor Sandra: you worked your magic to make sure it all came together and made sense. Thank you.

And finally a big thanks to my whole family for all their love and support during this project. Especially to my mum and my sister Anna for being my driving force, always pushing me on and asking, 'Why aren't you working on the book?': thank you. I would probably still be thinking about it without you. To my darling little girls Annabelle and Gabrielle, thank you for being so patient while Mummy was busy writing. I love you. And of course my husband Mat, for being a fantastic partner, advocate and case study, and enabling me to do this.

Introduction

If you are a dairy farmer who wants to significantly reduce your work hours, have more time to spend focusing on the business, cut costs in tight times and have a more efficient, profitable and sustainable farm, then this is the book for you. Well done for getting a copy — you are now beginning a journey towards transforming your dairy farm and business into a Lean Farm: making it a better, smarter, more productive farm using new thinking and ways of working.

After developing and rolling out the first comprehensive LeanFarm training program designed specifically for dairy farmers to more than 100 farmers across New Zealand in 2017, it was clear to me that there was a need for Lean in farming. My husband Mat and I realised how valuable these simple tools and principles were to any size of farming operation — from husband-and-wife teams to larger corporate farms.

In response to this demand I collated the training program into this book so that all farmers across New Zealand and the world can learn how Lean can significantly improve their dairy farm.

How did this book come about?

In 2013 Mat returned to his 1000-head family dairy farm in New Zealand. Dairy farming was a new industry for me and it struck me immediately that it lacked Lean thinking. The industry was far behind industries that used Lean management, and I knew our dairy farm would benefit significantly from applying Lean practices. As a result, we started implementing Lean principles on our own family farm.

Our two 'city' dogs love their new farm life in New Zealand

My degree in engineering led me to start my career at Toyota in Australia before spending a few years working with Toyota Motor Europe learning the insides of the Toyota Production System (TPS). In 2009, I moved into consulting and began teaching companies about the TPS, more widely known as 'Lean'. I have since worked with dozens of companies globally to implement Lean methodologies successfully.

Many might ask how Lean can apply to a dairy farm when it's completely different from making cars. Actually, dairy farming is essentially a primary industry production process. It converts inputs such as grass through cows into the output of milk, the end product that is shipped to the customer. Lean helps convert business inputs such as people, machines and materials, through value-adding processes, into the right outputs as effectively and efficiently as possible.

Lean thinking can significantly improve many of the problems that dairy farmers face every day. Long work hours, high staff turnover, repeat problems, breakdowns, wastage, safety and high costs are all factors that cause unnecessary stress for farmers. Learning how to see waste, work more efficiently, create engaged teams, be more productive and solve problems effectively will save time and cost, improve business performance and create a better farm.

With its origins in Toyota in the 1940s, Lean management has been used for a long time, initially in manufacturing industries, and over the past twenty or so years it has spread to just about every other industry sector—from defence, to healthcare, to finance. Today, Lean is a prevalent and proven management approach used by thousands of businesses globally, from start-ups to large corporates. It aims to create and deliver products or services that the customer values as effectively and efficiently as possible by eliminating all waste and striving for perfection through continuous improvement.

Our team (from left): Richard, Hannah, Mat, Anthony, Samii, Rob, Joey and Jaya

How will this book help you?

The Lean Dairy Farm details the key areas on a farm where Lean will make a substantial impact. It aims to provide farmers with the practical skills they need to apply Lean tools and thinking to their business to achieve sustainable and tangible results. Lean is an expansive topic. In this book I bring together and focus on the most relevant and valuable Lean tools and practices and explain them simply and with real farm stories. I will guide you through a step-by-step approach for implementing the key Lean tools you need to truly transform your farm.

Throughout the book I have also included examples of templates from our farm that you can use. All of these templates and more can be purchased in editable formats online at www.leanfarm.nz to save you time recreating your own.

Our own farm is also on this Lean journey. In this book, we use our farm as a case study and share with you not only our stories, but also real farm examples and activities to help you implement the tools in an easy, sustainable way and, importantly, bring your team along with you to get real, tangible results. Is our farm Lean yet? Is it perfect? Absolutely not! But we have started the journey and are continuously improving, while creating the right culture.

This book is broken into three key parts:

- Part I introduces Lean and why it applies to farming.
- Part II introduces 10 practical Lean tools that will help your farm eliminate waste and become more efficient.
- Part III focuses on how to create the right culture on the farm to embrace Lean thinking and an ongoing improvement mindset.

The 10 Lean tools I introduce in part II are:

1. *the 8 wastes*—'see' and eliminate the waste in your business
2. *5S workplace organisation*—save time and reduce waste by organising your business efficiently and effectively
3. *visual management*—create a visual business that achieves results
4. *standard work*—optimise processes to save time and make work easier and more efficient
5. *value stream mapping*—understand how things are *really* done on a farm and start to eliminate waste
6. *practical problem solving*—create a team of excellent problem solvers who can solve problems for good
7. *built-in quality*—create a culture on the farm where everyone does things 'right first time'
8. *total productive maintenance*—stop constant firefighting and unplanned breakdowns

9. *creating flow*—streamline your business process flow and improve productivity

10. *visual planning*—achieve business goals and targets effectively through improved planning.

Come and join me on the journey to making your farm a better business.

Myself, Mat and our daughters, Annabelle and Gabrielle, among our cows

PART I
The concept of 'Lean'

I have dedicated the first part of this book to giving farmers a little bit of a background and insight into Lean management so that you have a better understanding of what it is and where it's come from. For many of you this may be the first time you have heard of 'Lean' and I want to make sure you don't think it's some kind of exercise program! Also, it's important for you to understand the methodology, how it was developed and where it has come from so that when you are talking to people or your team about Lean you can confidently explain what it actually means and give it some context. It also gives you some confidence that this is a *real*, proven methodology that's used by thousands of businesses globally (including farms across Europe and the United States) and isn't something I've invented. I have also used this section to give you a brief insight into why Lean applies to dairy farming and why it can have such a big impact on your farm.

Chapter 1

Before we get started ...

The Lean Dairy Farm focuses on improvements that any farmer can make on their farm. However, to implement any type of improvement you will need to first be open to change, and to accept that your farm isn't perfect and there are opportunities to improve. Further, you must acknowledge that while there are some things that affect your farm that you can't control, there are many opportunities and improvements that are within your control and solely dependent on your decisions and management. In this chapter I will discuss two concepts to help you transition your thinking so that Lean can be applied successfully on your farm: 'external vs internal locus of control' and 'the "possible" mindset'.

External vs internal locus of control

Profit and success are based on a lot of factors. To truly make your farm a Lean Farm will mean thinking differently (see figure 1.1, overleaf), whether you are the farm owner, manager, assistant or anyone else working in the industry. First, it's important to note that Lean is not going to solve all your problems. It can't change many factors that are inherent in farming—what I call the 'external factors'. These include:

- weather
- global dairy prices
- interest rates

- politics
- taxes
- policies/regulations.

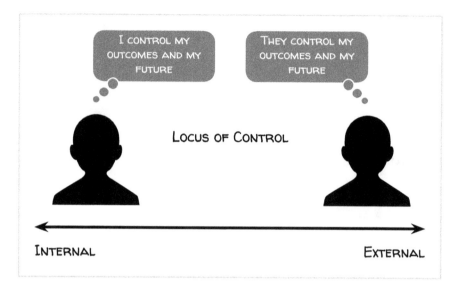

Figure 1.1: the locus of control
Based on a model by Julian Rotter

What Lean can help with is the things you *can* control on your farm. These are the 'internal factors' and include:

- your work environment
- what you do and how you do it
- your quality
- your waste
- your efficiency.

Therefore, to gain real value from this book and be able to implement Lean thinking effectively, we need to shift our mindset. We need to stop focusing and complaining about the external factors that we can't control. We need to stop believing that everything is outside of our control and blaming these external factors for our inefficiencies or telling ourselves things like 'we can't do anything', 'our hands are tied because milk prices are down' and so on. Instead we need to start accepting responsibility and taking control of the internal factors in our farm business.

This is what Lean and the concept of *kaizen* (we talk about this in chapter 2) are based on: continuously improving the things that are in our scope of control.

Toyota is a perfect example: they have many external factors that affect their business and are beyond their control, such as strong competition, fluctuating currencies, political situations, steel prices and so on. Yet they are still the world's most profitable car manufacturer because they use Lean principles and focus on the internal factors that they control and can continuously improve on.

The 'possible' mindset

The second concept that farmers need to open up to is something I call the 'possible' mindset (see figure 1.2).

Basically, we need to shift our thinking from 'not possible' mode to 'possible' mode. Instead of believing that it's not possible to do something, we need to start asking ourselves how we can make it possible: how can we do this?

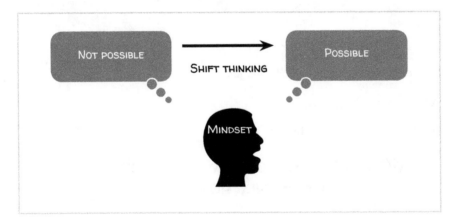

Figure 1.2: the possible mindset

It is believed that Henry Ford once said that if he had asked customers what they wanted, they would have said faster horses. Instead he invented the Model T Ford and the mass production assembly line, revolutionising society by providing the average American a car they could afford.

A team effort

Anyone know who these lads are?

Source: © Marco Iacobucci EPP / Shutterstock

What's so good about them?...

Did I hear you say that they are the best team in the world? Really?

WE CAN'T CONTROL HOW THE OTHER TEAM PLAYS — ONLY HOW WE PLAY.

So why are they so good? What sets them apart from every other team? They don't have different technology or special props. On the field they have the same rules, same weather, same number of players, same time. So what do they do differently to be the best? Well, there are probably three key things that set them apart:

1. *culture* — they have a very strong culture within the team, based on teamwork and respect

2. *drive* — from an early age, almost every Kiwi kid wants to grow up to be an All Black. Therefore, the fortunate few who get there are so

6

privileged to be part of the team that they have a huge motivation and ambition to do whatever it takes to stay on the team

3. *process*—this is the real key. This isn't just a bunch of random guys doing random things on the field that happen to make them win. No. They have a very clear plan of attack and a well-thought-out process that drives the right results. You'll often hear these players talk about the 'process' — about pulling themselves back to the process and concentrating on the task at hand. They are masters of process and continuously improving that process in small steps to make them better and better each time.

These are simple principles that we can take back to our farm by:

- creating an excellent culture within our farm business
- making people want to be part of our farm
- being masters of process and continuous improvement.

Mat puts our people and our cows before anything else

Hopefully you have switched your mindset to 'I am taking control of my destiny' and 'Everything is possible'. Now that you're mentally prepared for the Lean method, it's time to learn more about the concept of Lean and why it's so important to incorporate Lean into your farm.

Chapter 2

What is 'Lean'?

Lean is sometimes portrayed as complicated and very technical. When I listen to 'Lean experts' who haven't worked at Toyota but teach or write about Lean, I often feel they over-complicate things. Yes, Lean is a system, it is a philosophy, but it is based on some very simple notions that all businesses try to strive for in one way or another: produce products with a high-quality, low-cost, quick lead time, that satisfy customers and make a profit for the business. It is in the 'how' you do this that Lean helps your business: it is the system you use to achieve this goal, and this system is based on continuous improvement (an important concept that we discuss in this chapter) by eliminating all waste from your processes to help you be as efficient and effective as possible. In this chapter I will explain Lean in simple terms because it doesn't need to be complicated.

What exactly do we mean by 'Lean'?

There are a million and one different definitions of 'Lean' floating around these days. Here is what I have come up with:

> Lean is a best practice holistic management philosophy for any business that puts the customer at the heart and helps a business to achieve its strategy, purpose and vision as efficiently and effectively as possible, while being profitable. The aim is to provide customers with the products or services they want, when they want them and to the highest quality and lowest cost. This is achieved through systematically identifying and eliminating all *waste*, building continuous improvement into the DNA of the business and relentlessly pursuing perfection.

9

Regardless of which definition you use, what is important to understand is that *Lean should be a business strategy, not a one-off activity!* Lean is a holistic culture, philosophy, mindset and way of life for a business. Lean is a culture that focuses on:

- continuous improvement
- open-minded thinking
- challenging the norm
- the flow of value
- the relentless elimination of all waste
- a pursuit of perfection.

It gets us to think hard about what we do and how we do it and makes us ask ourselves 'Why?'

Lean is not about being 'mean' or trying to speed up processes, and it most certainly doesn't aim to increase productivity or efficiency at the expense of quality, safety, animal wellbeing, employee satisfaction, customer satisfaction or any other important factor. Lean focuses on process, and performance of the process. It strives for simplicity and common sense.

Furthermore, the biggest challenge when trying to become a Lean business is not the technical things. It is related to *people*. This is because it is easier to teach people technical stuff such as how to milk cows, put up fences or maintain a tractor; trying to change people's attitude and behaviour is much harder.

Figure 2.1 illustrates Lean in a nutshell (and this book is about making Lean simple).

One of the most important things I learned during my days at Toyota is **'doing the right thing will get the right result'**.

I have modified this a bit to

'DOING THE RIGHT THING, THE RIGHT WAY, WILL GET THE RIGHT RESULT'.

In other words, if you focus on the right things in your farm business, and everyone in your team does the right thing the right way every time (this means doing something right the first time (not taking shortcuts) and ensuring 100 per cent good quality), then you and your farm business will inherently get the right result.

Lean in a nutshell

CONTINUOUS IMPROVEMENT

+

PURSUIT OF PERFECTION

=

ELIMINATED WASTE	IMPROVED PRODUCTIVITY	REDUCED TIME
IMPROVED QUALITY	REDUCED COST	IMPROVED CUSTOMER SATISFACTION

=

A PROFITABLE BUSINESS

SIMPLE + COMMON SENSE

Figure 2.1: simplifying the concept of Lean

Lean Farm team activity

With your team, discuss the following:

— Has anyone heard of the term 'Lean'?

— Who has had previous experience with Lean (perhaps in other jobs)?

— Ask each person in your team to google the term 'Lean manufacturing', write down one sentence about what they discover and present their result to the team.

— With your team, watch the light-hearted video about Lean called 'Lean gone Lego' on YouTube.

— How can Lean thinking help your farm and team?

The history of Lean

Lean manufacturing is most certainly not new; it has been around for decades and is still gaining interest and being applied to different industries. This shows how powerful and successful Lean thinking is—if it weren't, it would have been just another fad that quickly lost momentum. It most certainly isn't something I have just made up! I want to share some of the background to Lean so that you can be assured that it is a very proven, beneficial methodology that has a long, successful history in a wide variety of industries.

How did Japanese manufacturing become a leader?

Japan went from bankruptcy to being one of the richest countries in the world. How? They started planting the seeds of what is now known as Lean:

- Japanese management realised the power of employee involvement.

- They made a commitment to continuous improvement.
- They defined the flow of value and ensured everyone focused on eliminating waste bit by bit.

In the 1940s, Toyota was only producing around 100 000 cars annually and was at a critical point financially during a difficult postwar period in Japan. Meanwhile, General Motors was a giant and successful automotive manufacturer producing in excess of 3.6 million vehicles annually. By 2015, roles had reversed: Toyota had increased its production a hundred-fold to more than 10 million annually to become the world's number-one automotive company, while GM increased its production over the same period to around 9.8 million vehicles (threefold) but went into bankruptcy and had to be bailed out by the US government. Interesting, isn't it? How did this happen? Toyota transformed by using Lean thinking (see figure 2.2).

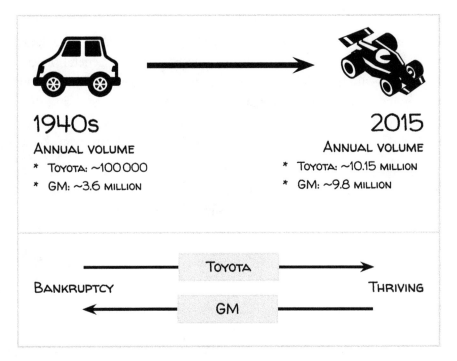

Figure 2.2: how Toyota transformed by using Lean thinking

A historical timeline of Lean

Lean has developed over a period of more than 100 years. Figure 2.3 depicts a rough timeline of how Lean evolved.

Lean thinking has some of its roots in what's known as 'scientific management', which was a theory of management established in the late nineteenth century by Frederick Taylor to analyse workflows and improve productivity using concepts such as standardisation. Henry Ford further refined the concepts from the scientific management era with his mass-production standardised assembly lines. Japanese engineer Taiichi Ohno visited the United States to learn about what Henry Ford was doing and how supermarkets created flow of product 'Just In Time' by restocking their shelves based on customer demand. Based on his research, Taiichi Ohno developed the Toyota Production System (TPS), which became one of the world's best management philosophies for eliminating waste and creating a profitable, high-quality, flexible business.

In the late 1980s, authors James P Womack, Daniel Jones and Daniel Roos studied the three largest automotive producers—Toyota, General Motors and Ford—and their differences. They spent time in Japan to understand the approach Toyota and other Japanese companies were using and they published a book about their research called *The Machine that Changed the World* in 1990. This book essentially introduced the term 'Lean' to the world by describing the TPS as 'Lean production'.

Today, Lean is a management approach used by almost every industry sector and any type or size of business globally. It is a proven method for eliminating waste and creating successful, profitable, efficient businesses.

The evolution of Lean

1920s+

* FORDISM
* MASS PRODUCTION
* STANDARDISATION

1950s

* TOYOTA'S TAIICHI OHNO VISITS US CAR MANUFACTURERS AND SUPERMARKETS
* INTRODUCES THE TOYOTA PRODUCTION SYSTEM (TPS)

1980s

* TPS INTRODUCED IN USA WITH A TOYOTA/GM JOINT VENTURE
* TQM (TOTAL QUALITY MANAGEMENT) MOVEMENT

1990s

* SIX SIGMA GAINS POPULARITY

2010s

* LEANFARM—LEAN IN FARMING

1880s

* SCIENTIFIC MANAGEMENT (TAYLORISM)
* PROCESS MANAGEMENT

1940s

* TOYOTA NEAR BANKRUPTCY
* POST–WAR JAPAN
* ISO ESTABLISHED

1970s

* TPS MOVES TO OTHER MANUFACTURING SECTORS

1990

* THE TERM 'LEAN' INTRODUCED TO THE WORLD IN THE BOOK ABOUT TOYOTA CALLED *THE MACHINE THAT CHANGED THE WORLD*

2000s

* LEAN INCREASES IN POPULARITY IN OTHER INDUSTRY SECTORS

Figure 2.3: the evolution of Lean

A holistic approach

Lean is more than just a one-off improvement: it is a holistic approach to your business. To be a truly Lean business, you need to apply Lean thinking across all elements of your business. It needs to become just 'the way we do things around here' and part of the DNA of your business. Almost every business these days is made up of three key components—process, people and technology—and is underpinned by *purpose*. Lean must be applied to each one of these three components to be effectively integrated into the business.

Mat checking the condition of the cows to ensure they are healthy (high-quality)
Source: Grant Matthew

Figure 2.4 is my vision of a Lean Farm based on the TPS house. A business must make a *profit* to exist. In a Lean world this is done by improving productivity. To improve productivity, Lean methodology and tools are applied to people, process and technology. This is supported by a Lean culture that is based on continuous improvement and respect.

Ultimately, it is the culture of a business that will drive the right Lean thinking and principles to achieve the desired outcomes.

FIX THE PROCESS FIRST; THEN ADD TECHNOLOGY!

I will also mention here something about technology. Technology is great and farming has had significant progress and transformation over the past century due to evolving technology. However, we must never assume that technology is going to solve all our problems. If you have poor processes, or no processes, in place and you invest a lot of money in technology thinking it is going to solve all your problems, you might be disappointed. It is always important to create a standard, reliable foundation before slapping some technology on top of it. Otherwise you might find that you have put a great piece of technology on top of bad processes and systems and you don't see the full benefit of your investment. Successful Lean companies such as Toyota always look at their processes first and refine and improve them. Once they know they have good, reliable, repeatable processes, they consider adding some new technology to further improve them. They don't just put in technology for the sake of it. They know it's important to fix the process first, then add technology.

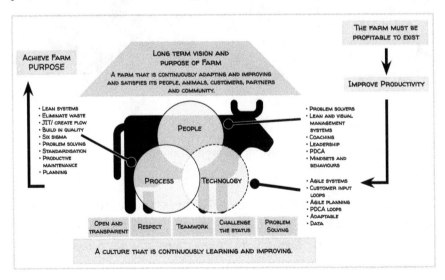

Figure 2.4: Jana's Lean Farm

Who uses Lean?

The short answer: *everyone*!

As I said earlier, this isn't something I have just made up and pulled out of my back pocket. Nor is it just some fad. Over the past 20 years in particular, almost all industries have introduced some level of Lean management. And many businesses, from global organisations to small businesses, are introducing Lean concepts today. This shows that while the methodology has been around for almost seventy years, it is proven, it works and it is absolutely still relevant today. There are numerous books out there about Lean for tech start-ups, large corporates and even small-scale market farmers.

The power of people involvement is a fundamental aspect of Lean philosophy

I have worked with all types and sizes of business and across most industry sectors, from mining, to rail, to healthcare, to banking. Lean is applicable to every one of them. Figure 2.5 demonstrates how Lean has spread to different industries over the past two decades.

It's proven and it works!

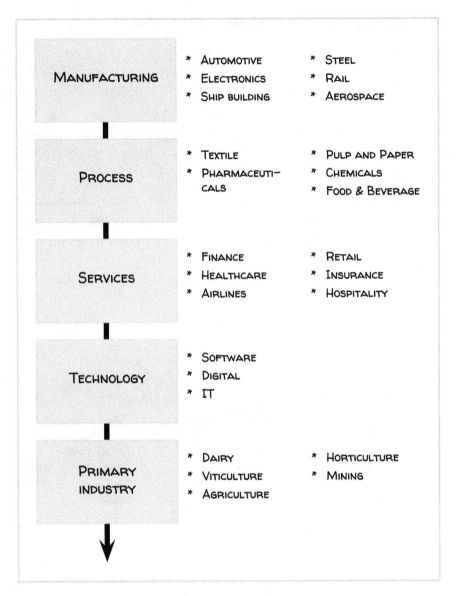

Figure 2.5: some of the many industry sectors that use Lean

Profit

The ultimate purpose of introducing Lean into your farm is to create a profitable business.

Why does your farm need to be profitable?

There will be many reasons why you want to make sure your farm is profitable: growth, stability, security, flexibility, investment in people, systems, infrastructure. The biggest reason, however, is that if your farm doesn't make money, you probably won't be in business very long. You need to be profitable to ensure a sustainable business that can operate in the long term. In other words, you need to make a profit in order to exist.

There are different ways you can increase your profits. Traditionally, people try to increase the prices of their product to increase their profit margin. Dairy farmers are similar — we hope like mad that milk prices go up so that our profit margin increases. Unfortunately, however, we are not in control of milk prices, so there is no guarantee that we will be profitable.

This approach is known as the 'cost plus principle' and is illustrated in figure 2.6.

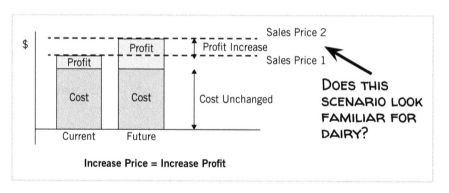

Figure 2.6: the cost plus principle

Lean uses the philosophy that you can't change factors outside of your control (these are the external factors, such as demand or milk prices), but you can control your internal factors, such as your internal farm costs. Therefore, to make a profit, you need to reduce your internal costs by improving your farm productivity. To improve productivity, you need to eliminate waste and continuously improve using Lean principles and thinking. This is known as the 'cost reduction principle' (see figure 2.7). It involves forgetting about the price of milk, which you can't change, and instead focusing on what you can do inside your business to reduce costs.

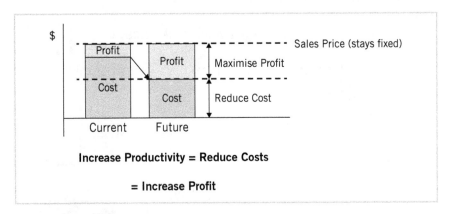

Figure 2.7: the cost reduction principle (Toyota/Lean approach)

Profit margins can be low for dairy farmers, particularly during tough times such as those we have seen recently. As farmers can't control price, cost reduction or milk production increase (without increasing costs) is our biggest opportunity to improve our profit margins. Understanding our costs, trying to cut out waste and improving our processes is essential for all farmers.

Recently I discovered a pertinent quote from a Toyota CEO that is applicable to dairy farmers—we should all think about cost in this way:

Cost reduction must be a fundamental part of our [business]. We must encourage all our members to have a cost conscious mind and ... lead and stimulate ... cost reduction ... [it] is the single biggest weapon to improve our profitability ... [and] our main defence against negative external factors.

Lean Farm team activity

With your team, discuss the following questions to give everyone an appreciation of profit and cost:

— Why does your farm have to be profitable?

— What happens to your profit when milk prices drop?

— How much control does your farm have over what milk price you get?

— How much control does your farm have over your internal costs?

— Why is it important to try to reduce your internal costs sustainably?

Rob laying water pipes at the new farm, Longmere

Continuous improvement

Lean is underpinned by a culture of continuous improvement. This is the fundamental philosophy in Lean businesses and forms part of the DNA of the business. Our farms should be continuously improving businesses. Why?

Competitiveness

Are our farms competitive in terms of cost, quality and responsiveness? Should we be?

Most of our milk is processed into whole milk powder, which is then traded on the global market. We therefore must participate in highly competitive world markets and absolutely need to be competitive. Unless we create a significantly appealing differentiator of our milk and can demand a premium for it or basically ask our price, we will continue competing with the rest of the world.

To ensure a competitive advantage and/or feasibility of our farms we must improve more quickly and be more cost effective than our competitors around the world (see figure 2.8, overleaf). To do this successfully, our farms need:

- a clear vision
- strong leadership
- everyone involved
- a high skill level
- clear and SMART (specific, measurable, achievable, relevant, time bound) performance targets
- Lean thinking
- a culture of continuous improvement.

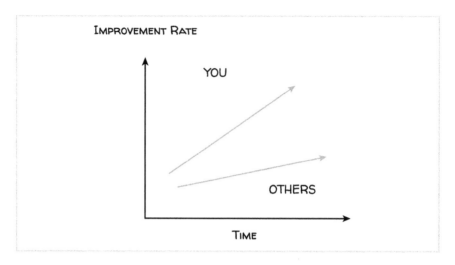

Figure 2.8: comparing improvement rates

Change

To continuously improve and stay competitive, we need to be able to change the right things in our businesses at the right time. Importantly, we need to accept that we can't continue doing things the way we have always done them. What worked in the past may not necessarily work in the future. This is because the dairy industry has changed considerably over the past few decades. We now have more public pressure, more regulations, different technology, different markets, more dairy farmers and larger scale farms. We need to adapt to our changing environments and consumer demands. One of my favourite quotes that depicts the importance of change is:

It is not the strongest of the species that survives, nor the most intelligent that survives. It is the one that is most adaptable to change.

From Darwin's Theory of Evolution

Embracing change: technology has dramatically changed how we do things on our farm. One example is the use of helicopters to apply fertiliser

So what is continuous improvement?

There are two key ways of improving your business: adding new technology or improving processes. Technology, as we know, can be expensive and time-consuming to implement. Process improvement, however, is much easier and cheaper to do and can also lead to significant benefits. Continuous improvement is focused on process. In the Lean Japanese world continuous improvement is known as *kaizen*.

SEEK SMALL IMPROVEMENTS IN PROCESSES
AND PRODUCTS SO TOMORROW IS BETTER THAN
TODAY — EVERY DAY, EVERYWHERE, EVERYONE!

Kaizen essentially means improvement in small steps. The philosophy is about integrating improvement into everyday work so that it is done daily by everyone rather than it being just a one-off project.

Lean Farm team activity

With your team, brainstorm the following:

— How do we improve right now?

— How do we involve our employees in improvement right now?

— What can we do better?

We should be constantly looking at how we can improve the way we do things to improve our quality, efficiency, innovation and bottom line, and to strengthen our competitive position. It is about working smarter, not harder.

Continuous improvement is *continuous* — that is, we should be thinking about it every day, every week, even every hour. No matter what type of work we are doing on the farm, we should be asking ourselves:

- What are we doing?
- How are we doing it?
- Why are we doing it?
- How can we do it better?
- Where is there *waste*?

Continuous improvement (CI) isn't just the responsibility of the owner or manager of the farm. It should be everyone's responsibility. Everyone should be responsible for challenging the way things are done and finding ways to improve and make tomorrow better than today (see figure 2.9). The leadership team on the farm should reinforce, encourage and reward a continuous improvement mindset among their teams.

Importantly, they need to ensure that the culture on the farm promotes sharing ideas, openness, seeing change as positive, and empowering teams and people to improve.

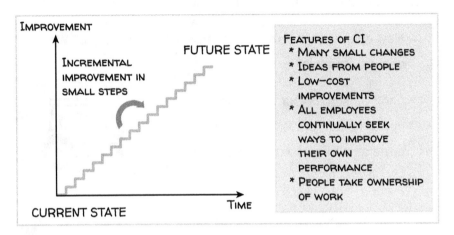

Figure 2.9: standardise, improve, then innovate

Shoichiro Toyoda, the grandson of Sakichi Toyoda (who was the founder of Toyota Industries and father of many of the Lean concepts used today) talked about the 3 Cs: Creativity, Challenge and Courage (see figure 2.10). I think these are three wonderful words to instil in the minds and attitudes of people in any business. It is these 3 Cs applied in a measured way that can truly make the impossible possible and turn a business from mediocre to great.

Figure 2.10: the 3 Cs of a successful business

Lean Farm team activity

With your team, brainstorm the following:

— How open is your farm to challenging the status quo? What can you improve?

— Each person in your team should identify one simple improvement that they can take ownership of and action themselves.

Two of our team—Jaya (left) and farm manager Joey—brainstorming the annual strategy

It won't happen overnight

When you start implementing some of the things I talk about in this book you will find you get frustrated that things are taking longer than you would like. This is *normal*. Don't be discouraged that you can't get everything done in one go. Instead, just try and do a little bit every day. This is continuous improvement. Remember: it is *continuous*. Just make one little improvement each day and you will slowly transform your farm into a Lean Farm. Importantly, if things are not going how you want them to first time round, try to stop yourself from giving up or reverting to your old habits. Just keep at it.

You should now have a good appreciation for what Lean is, where it comes from, where it's used and some of its benefits. But you may still be wondering how lean can apply to your farm. In the next chapter, I will explain why lean thinking is relevant to dairy farming, or any other farming, and the value it can bring to your business.

Chapter 3

Applying Lean to dairy farming

Why is Lean relevant to your farm? Well, let's look at the simple model shown in figure 3.1 (overleaf).

Lean is about converting business inputs such as people, machines, equipment, and materials through highly efficient and effective processes, into the right business outputs—products/services of high quality and in the right quantity—that are produced and delivered on time, at the lowest cost, resulting in satisfied customers and employees, and ensuring safe, animal-friendly and environmentally sensitive practices.

Lean helps you to create processes and operations that deliver the right results for your farm.

Efficiency vs effectiveness

We are all very, very busy farmers, right? But does being busy mean we are also effective and efficient? What is the difference between efficiency and effectiveness? Can we be one and not the other? Absolutely!

Efficiency is about how quickly you can do a job. Effectiveness is about how well you do the same job so that it delivers the desired outcome (what the customer or business wants).

When we look at our business processes, we need to ensure that we are creating the right balance between efficiency and effectiveness. If you think about your day honestly, for what percentage of your day do you think you are efficient?

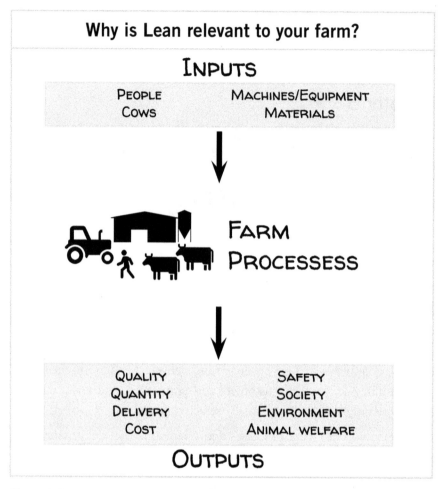

Figure 3.1: converting inputs into the right outputs effectively and efficiently

It is sometimes difficult for our team and even management to see why efficiency and effectiveness are equally important. Here are farm examples of how focusing on only one of these without the other will deliver a poor result for your farm.

A Lean Farm example

Efficiency vs effectiveness

Jo is on cups on today. He is milking incredibly 'efficiently': he is milking 450 cows in an hour in his rotary shed with his very quick method of flicking cups on with one hand. Unfortunately, because Jo is putting cups on so quickly and has the platform going at a fast speed, he hasn't got enough time to check for mastitis properly. Therefore, many cows with mastitis are being missed. This has resulted in a couple of poor quality grades lately. In other words, Jo is not being very 'effective' in producing high-quality milk.

The next day Bob is working at cups on. He turns the platform speed right down so that it takes him three hours to milk the same 450 cows. He spends a lot of time checking for mastitis and he has ensured perfect-quality milk. In this case Bob is being effective; however, he is not being very efficient.

Lean Farm team activity

Can you think of an example on your farm where you were effective but not efficient? What about efficient but not effective? Discuss this with your team.

What could you have done differently to be both effective and efficient?

The dairy farm business

Dairy farming is a unique type of business. It has many challenges that make it quite different from most other industries. At the same time, it also has many positives that make farming a little easier than some other industries.

Table 3.1 is most certainly not exhaustive. It simply highlights some of the more apparent positives and challenges dairy farmers face around the world.

Table 3.1: positives and challenges of dairy farming

Positives	Challenges
One product (not producing potentially hundreds of different products simultaneously)	Only one customer (you can't differentiate or sell more/get more customers)
One customer (simpler to meet one customer's constant needs than the varying needs of multiple customers)	Only one product (you can't change or add product variety easily to differentiate or diversify)
One set of milk specifications to meet	Commodity price reliant on global market (you can't dictate)

Positives	Challenges
Can push product out the door (not made to order/demand as this is left to the dairy company to manage)	Weather
Don't have to compete for customers	Interest rates
Generally guaranteed that customers will buy your milk	Economic landscape
Repetitive processes	Political landscape Environmental/social/welfare concerns

We all know that farming can be a tough job. Generally, a farmer has to wear several different hats: mechanic, crop expert, vet, HR and production manager, among others. Often, one person on the farm does the same amount of work as what some other businesses might need several people to do. Therefore, for farmers, time is precious. The more we can understand where we're spending our time — what type of activities are consuming our day — the more we can look for opportunities to improve what we do and save time and money.

If we look at our dairy farm processes more closely, we can essentially break them down into three cyclic processes that need to be managed simultaneously. These three processes essentially create the annual farm season, which is mostly the same every season. The brilliant thing about farming is that while the decision making might be different, the actual process is the same. Whether it is done daily, weekly, monthly or annually, most farm processes are repeated so we can learn from them and continue to develop, evolve and improve them. Figure 3.2 (overleaf) highlights the key processes.

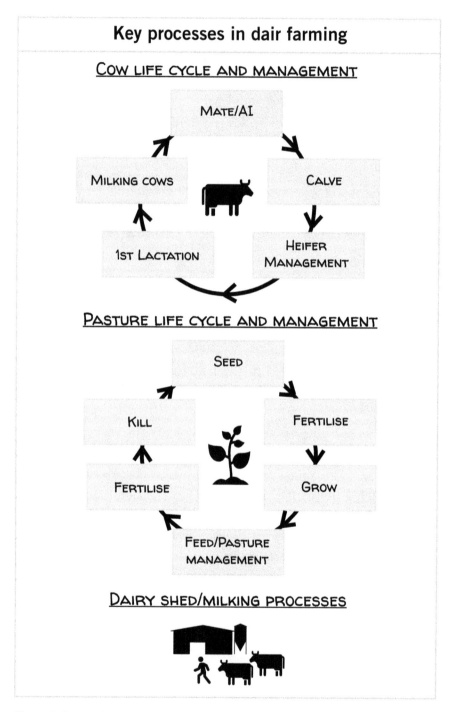

Figure 3.2: the key processes in dairy farming

The dairy farming season cycle

Let's have a look at the overall farming season cycle. Figure 3.3 shows a typical farming season. As you can see, the general cycle during a season is the same each year unless you decide to make key changes such as changing to both spring and autumn calving (split calving). But even then it will become a repetitive cycle once the change is made.

Again we are talking about a repetitive process: another great opportunity to standardise how we do things at each stage, identify opportunities to do things better and continuously improve every season.

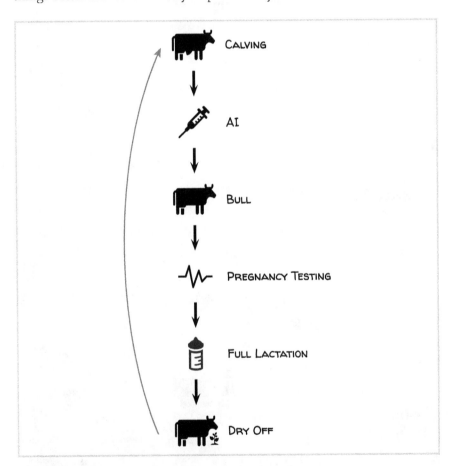

Figure 3.3: the typical dairy season cycle

Dairy farms are, in fact, the perfect environment for Lean thinking and it's a wonder that we aren't all using it already. Dairy farms have many elements that Lean can help with:

- equipment, facilities, machines
- product (milk)
- materials
- processes
- people
- physical flow of animals, machinery and people.

By looking at each of these elements objectively and analysing how they all work together, we can optimise our business operations and processes to make our farms much more productive.

Lean can also help us to think more about process design, equipment design, layout design and flow design so that whenever we have the opportunity we can actually design our farms to be Lean and incorporate efficiencies into the way we build and design our farms.

Lean Farm team activity

With your team, brainstorm and list at least 10 repetitive processes on your farm.

One of the rotary dairy sheds at Grassmere Dairy

Benefits of Lean for a farm

I can't even begin to list all the significant benefits Lean can offer your farming business, but here are a few:

- cost reduction
- waste reduction
- time savings
- better quality
- better animal health and results
- increased production
- reduced rework and mistakes.

Importantly, Lean will empower your team, giving them autonomy and the ability to make the right decisions and take control. This means farm owners or sharemilkers can step out of the day-to-day and spend time where it is needed most: working on the business and coaching and developing the team.

From a people point of view, there are also significant benefits to introducing Lean. Lean will help your team by:

- making work easier
- creating a more consistent workload
- creating a safer work environment
- reducing stress and frustration
- reducing work hours
- empowering them
- improving job satisfaction.

Here are some examples of tangible results that Lean has brought to some of the businesses I have worked with:

- a 55 per cent reduction in hospital admission waiting time
- a $4.3 million reduction in the cost of retail banking transactions
- a 33 per cent reduction in maintenance time in defence industry
- a 67 per cent reduction in the quality defects of engine components

- a 54 per cent reduction in process lead time for the assembly of a electrical component
- a 58 per cent increase in capacity of an energy provider.

On our farms there are endless opportunities for seeing some big, tangible wins by introducing Lean, including improved production results, cell count and overall quality, and even optimised feed conversion. You can see many of these in figure 3.4.

Performing a simple, everyday process such as feeding out inefficiently can result in lost feed, fuel and time, costing you money
Source: Grant Matthew

All of this ensures we are competitive, profitable and sustainable as a business.

Lean Farm team activity

With your team, brainstorm and list at least 10 benefits that you think Lean will have for your farm.

So there you have it—you now have a better understanding of what Lean is and how it can apply to farming. Can you imagine the enormous positive impact that adopting Lean will have on your dairy farm?

Benefits of Lean and continuous improvement

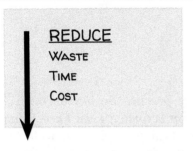

REDUCE
WASTE
TIME
COST

INCREASE
QUALITY
CUSTOMER EXPERIENCE
EMPLOYEE OWNERSHIP
AND MOTIVATION

Customer benefits

* Faster on-time production
* Higher quality product
* More product
* Lower costs may translate to improved value
* Better engaged staff = better customer service
* Better customer satisfaction

Business benefits

* Increases people's engagement
* Helps meet business targets (cost, quality, service)
* Encourages knowledge sharing across teams
* Reduces staff turnover
* Encourages people to question the status quo and come up with better ways of doing things (innovation)
* Time to focus on value add activities
* Improved quality = less rework
* Increased profit

People benefits

* Increases job satisfaction: we have a say in how things operate
* Removes frustrating elements of work
* Empowers people to take action to improve the business
* Making work more efficient helps support improved work–life balance
* Better place to work

Animal benefits

* Happier cows
* Healthier cows
* Correct animal handling with better standardisation/processes
* Better animal care
* Prompt/accurate diagnosis and treatment (standards and right tools/materials at point of use)

Figure 3.4: the benefits of Lean and continuous improvement on a farm

In part II I put Lean thinking into practice by introducing you to 10 Lean tools that will help your farm become a Lean Farm. Many of these tools are very simple and practical to implement, while others are more challenging and require a change of mindset and behaviours. Introducing these tools on your farm will help you achieve an effective, efficient farm operation with minimal waste, while building a highly engaged team.

10 Lean tools that will transform your farm

Part II is essentially the guts of this book. It explains how you will put Lean thinking into practice on your farm by implementing 10 Lean tools and methodologies. I will call them 'tools', but most of the concepts I talk about in part II are more than just tools. They are fundamental principles of Lean that need a business-wide mindset change. Principles such as 'building in quality', 'visual management' and 'problem solving' are implemented using various specific tools, but they are actually much bigger than that. Ideally, they should form part of the DNA of your farm and your strategy—that is, 'how we do things around here'. They should be integrated into your farm's management approach so that they are part of everything you do in a holistic and sustainable way. Of course, this won't happen overnight. The idea is that you start to slowly implement some of these tools, one step at a time, and begin to incorporate them into your way of working. Most of them are simple, commonsense concepts that all of us can do; it just takes conscious effort, focus and commitment to make them happen and once you do you will reap huge rewards.

Chapter 4

The 8 wastes on the farm

Lean is the relentless pursuit of *identifying and eliminating waste* in all of its forms in order to improve business performance and customer satisfaction.

So what is this *waste*?

The three components of work

To understand what waste is we first need to understand our work and the meaning of *value*. Everything we do (that is, all our tasks, activities and processes) can be broken down into three key components (as shown in figure 4.1, overleaf):

- value
- incidental work
- waste.

Value (VA)

Value (also known as value-added work or VA) is any activity we do that the customer sees as beneficial and would be willing to pay for. Value is created by any activity that changes the fit, form or function of a product or service provided it is done right the first time. It is important to note that value is seen through the customer's eyes so it is 'value in the eyes of the customer', *not* value in the eyes of the business.

The three components of work on a farm

VALUE
ACTIONS THAT THE CUSTOMER
VALUES AND WOULD PAY FOR
* THE CUSTOMER CARES ABOUT THE ACTION
* THE ACTIVITY CHANGES THE FIT, FORM OR
  FUNCTION OF THE PRODUCT

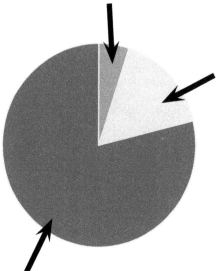

INCIDENTAL WORK
ACTIONS AREN'T OF VALUE
TO THE CUSTOMER BUT ARE
NEEDED OR POSSIBLY ENABLE
VALUE–ADDED WORK
* ACTIVITIES THAT ARE REQUIRED
  TO CONDUCT BUSINESS
* REGULATIONS/LEGAL
  REQUIREMENTS/COMPLIANCE

WASTE
ACTIONS THAT ARE NOT CONSIDERED OF
VALUE TO THE CUSTOMER
* TASKS, MATERIALS, RESOURCES THAT ARE NOT
  NEEDED TO MEET CUSTOMER REQUIREMENTS
* ACTIONS THAT ADD TO COST BUT HAVE NO VALUE

Figure 4.1: the three components of work on a farm

The most obvious example of value would be the process of putting cups on to milk cows, which creates a product—milk (or more accurately, extracting milk from the cows to the consumer).

Every time products and services are handled something is added to its cost, but not necessarily to its value.

Henry Royce, co-founder of Rolls-Royce

To work out whether or not an activity is a value-added step, ask yourself:

- Does the customer care about it?
- Does it physically change the product?
- Is it done right the first time?

Lean Farm team activity

What are some examples of value-added activities on your farm? List at least five and discuss them with your team.

Incidental work

Incidental work—also known as 'non–value added but necessary', or business value added—is activities that don't necessarily add any value to the customer but that have to be done. Examples of incidental work are activities necessary due to policy, legislation, compliance or similar. This could be the need to complete safety or quality compliance documentation to meet certain safety or hygiene industry standards. Another example could be doing regular herd testing—while it might be helpful for the business or industry to collect data, most likely our end consumer couldn't care less.

Waste

And finally there is waste (also known as non–value added work). This is everything else that we do besides value and incidental work.

It does not add any value to the farm or customer but does add cost to your business. Waste is any work, materials or resources that are beyond what is needed to meet customer requirements.

Finding waste in our farm processes can give us a lot of opportunities to improve.

Lean Farm team activity

How much waste is on your farm? Ask each member of your team to estimate the percentage of waste in their processes. Discuss this.

A closer look at waste

So how much waste do you think there is in our everyday work on a farm? Would you believe it's often 95 per cent? Yes, that's right ...

95% OF WHAT WE DO IS WASTE!

But don't worry! There is a reason for this ...

If we look at every single step taken to complete a process from start to finish and assess each step from the customer's point of view—and if it adds value—in almost every process I have studied the waste is more than 95 per cent.

In any end-to-end process—whether it be drenching cows, feeding calves or maybe doing a plant start-up—we carry out some 'waste' tasks, then some value-added tasks, followed by some more waste, then a bit of value added, then more waste again and so on. So if we recorded every step we took in our process, asked whether it is value or waste and then drew our process in terms of components of value-added tasks and waste tasks we would end up with a diagram similar to figure 4.2.

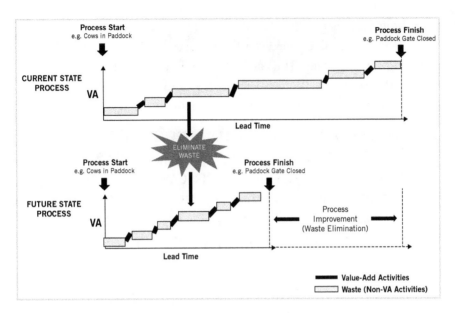

Figure 4.2: our current and future state for the end-to-end process of milking

What we can see immediately is that the biggest chunk of our process is actually the shaded tasks—in other words, waste. Traditionally, businesses don't see or think about waste so they spend their energy on improving the tiny, value-added process steps. Often, improving value-added steps such as introducing technology and automation is expensive or complex. Businesses then wonder why, after making some improvements, they haven't really gained significant time savings in terms of the overall process.

So what we want to do instead is really challenge these 'waste' tasks and start to eliminate or reduce them, rather than focus our energy on the black (or value-added) bits in figure 4.2, which only contribute to a small portion of our time in the end-to-end process. By doing this, we can create a much bigger impact on reducing our end-to-end process time compared with only focusing on the small number of value-add tasks. Eliminating waste steps is also usually cheaper, easier and we get a bigger 'bang for our buck'.

Figure 4.3 (overleaf) is a version of figure 4.2 drawn another way.

The first bar in figure 4.3 indicates a typical business, whether it be a farm or a café. The majority of the process time from end to end is waste in the eyes of the customer. Because the VA activities we do are the most visual and understood, we can easily see, measure and improve these. They therefore become the focus for improvement. On a farm this could be the cups-on process as this is probably the most value-added process and usually where we focus our time and energy on improving. This can be seen in the second bar in figure 4.3: we improve the value-added component of our end-to-end process, which is only a small time component of our overall end-to-end process.

THE LEAN APPROACH TO IMPROVEMENT IS TO FORGET THE VALUE–ADDED ACTIVITIES *FOR THE TIME BEING* (NOT FOR GOOD). INSTEAD, FOCUS ON ELIMINATING ALL THE WASTE FROM THE PROCESS — THIS IS WHAT'S CONSUMING MOST OF OUR TIME!

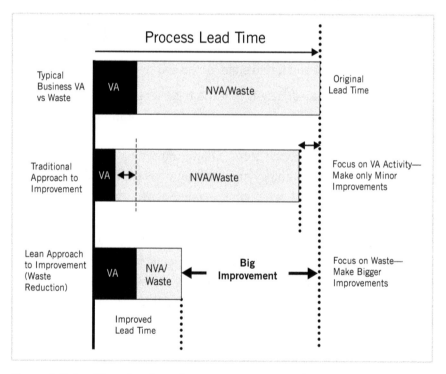

Figure 4.3: traditional vs Lean improvement approach

If we focus on the waste component of our work we will generate the biggest saving in time and cost. Once we have eliminated or reduced as much waste as possible from the process, we can then start to look at the value-added components and further refine these.

A Lean Farm example

VA (value) and waste

Imagine putting a red dot on one cow in our herd and following the cow through every single step of the morning milking process from when the cow leaves the paddock until the cow is back in another paddock. If we timed each step from walking, to waiting, to being milked, to walking back, we would see that the time spent milking the cow (perhaps eight minutes) is a very small proportion of the total time of this process for the cow (and in the eyes of the customer), which could be three hours from start to finish. Essentially, the cow is spending only eight minutes out of three hours being milked (if there is no time feeding on the feed pad) rather than being in a paddock doing something value adding—eating grass.

Therefore, if we spend all our time and energy trying to improve the eight minutes of our three-hour process, and make our milking six minutes, then we haven't really made much of an impact on the duration of the entire process for the cow, have we? (Of course, it will reduce the time for the person in the shed, which is good, but remember here we are looking at the process from the point of view of the customer and each cow.)

What we want to do instead is focus on the other two hours and 52 minutes of our process and try to work out what is waste in this process and how we can improve it to significantly reduce the time. This could be the time the cow is waiting on the yard, or the time it spends walking to the shed.

| COW LEAVES PADDOCK | 1 HOUR 30 MIN | MILKING 8 MIN | 1 HOUR 20 MIN | COW BACK IN PADDOCK |

Can you imagine all the time you would have if you could eliminate 95 per cent of what you do?

Unfortunately, it's easier said than done. Waste is very hard to see if we don't know what to look for. Waste is integrated into everything we do and exists in all parts of our farm business. To be able to find waste in our processes, we need to critically analyse every step in every process. This can be very hard for farmers because we are so used to doing things the way we have always done them. This obscures our vision and we don't scrutinise and question the processes around us enough to identify whether we really need to be doing something or whether it is in fact waste. Figure 4.4 explains this.

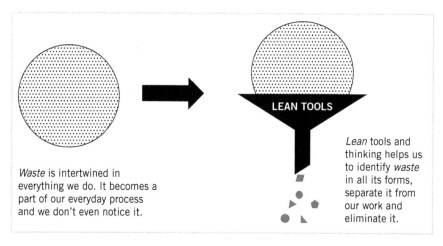

Waste is intertwined in everything we do. It becomes a part of our everyday process and we don't even notice it.

LEAN TOOLS

Lean tools and thinking helps us to identify waste in all its forms, separate it from our work and eliminate it.

Figure 4.4: identifying waste

As a result, because waste is hard to see, we often spend all our time and energy improving the value-added processes (which are easy to see) instead of focusing on the 'big elephant in the room': the waste.

Normal waste and bad waste

I was delivering my LeanFarm training module on waste to a group of farmers and one asked me if all waste is bad waste. Isn't there 'normal' waste and 'bad waste'? My response was 'NO!'

A team member driving a tractor between sheds

All waste is bad and it is important to look at waste as bad. If we start to accept waste in our farm processes, we won't ever be able to improve. While all waste is bad and we should strive to get rid of it, there will be waste that we might not be able to eliminate *at this point in time* (potentially due to our landscape or infrastructure). But this doesn't mean we shouldn't be aware of it and strive to eliminate it in the future through continuous improvement.

What is all this waste?

You are probably asking right now, 'If we have 95 per cent waste in our process, then what is all this waste?' At Toyota there are three key types of waste that also apply on the farm: Mura, Muri and Muda. We call these the 3 Ms of waste (see figure 4.5, overleaf).

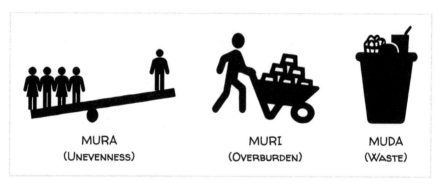

MURA	MURI	MUDA
(Unevenness)	(Overburden)	(Waste)

Figure 4.5: the 3 Ms of waste

Mura: unevenness

'Mura' is the Japanese word for 'unevenness'. This is work that is uneven in demand or workload, or that is inconsistent. This usually appears as a work imbalance between team members, time fluctuations or variations in work content.

Farm examples of Mura include:

- *the calving period*—peak workload during a few weeks that is not balanced
- *the maize or crop harvest*—high peak workload for crop managers or contractors during a certain period of the season when they are potentially working through the night to get maize cut and then have a quiet period
- *milk production*—milk volumes fluctuate during the season from no milk during dry-off to peak during full lactation. This unevenness in milk production volumes results in peak milk prices and off-peak milk prices.

Unevenness adds to chaos on our farms as well as costs in having to rely on relief labour or overtime. Ideally, wherever possible we should try to distribute our activities evenly through good planning, standardisation and better processes.

Cutting maize for silage on our farm

Muri: overburden

'Muri' is the Japanese word for 'overburden'. This is where a task that someone is doing is unnecessarily difficult and creates a burden for the person or process—in other words, a person or machine working beyond their natural limitations. This type of waste can result in accidents, safety concerns, team absences and frequent breakdowns. Overburden can be in three forms:

- *mental*—excessive decision making, stress, excessive overtime, poor training
- *physical*—excessive lifting, bending, walking, stretching
- *machine*—not working properly or working beyond its capability due to design, layout, use and maintenance.

Farm examples of Muri include:

- having to lift calves during calving
- stretching to reach control pads on the rotary platform
- bending to reach Vat taps that are in awkward positions
- walking back and forth in a herringbone dairy shed while milking
- towing a too-heavy load of calves behind the ranger and the tow bar breaking off.

On dairy farms, we have many examples of overburden. Unfortunately, this does make our farms less efficient, potentially unsafe and more costly. It's important to try to identify examples of overburden in our everyday work and remove it.

Lean Farm team activity

With your team, think about and discuss at least three examples from your farm of each of the following:

— unevenness

— overburden.

A team member bending to reach awkwardly positioned taps while doing a milk vat wash

Muda: waste

'Muda' is the Japanese word for 'wastefulness'. This comprises activities that are wasteful, unproductive and do not add value to the customer. We will spend most of this section discussing Muda (or waste), which can be broken down into what is known as 'the 8 wastes'.

The 8 wastes

The majority of waste (or Muda) in our processes (besides unevenness and overburden) can be categorised into eight areas known as 'the 8 wastes'. This is essentially a framework to help you start to think about and look out for the different types of waste in your farm processes. It also helps you to know where to focus your farm process improvements. (In the Toyota Production System there are only seven wastes, with the eighth (non-utilised people, intellect and resources) being added by the outside world as this was an area in which most companies did not perform well (farmers included!). Figure 4.6 illustrates the 8 wastes.

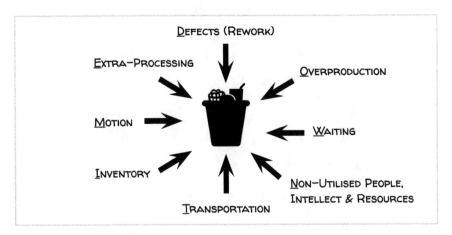

Figure 4.6: the 8 wastes (DOWNTIME)

DOWNTIME

An acronym for remembering the 8 wastes in business is DOWNTIME:

- Defects
- Overproduction
- Waiting
- Non-utilised people, intellect and resources
- Transportation

- Inventory
- Motion
- Extra-processing.

DOWNTIME means if you're spending time on work that involves any one of these wastes, it is effectively downtime for your business, adding cost and no value.

Walking is considered to be motion and is a waste

Let's now look at the 8 wastes in more detail.

Defects

'Milk went down the drain because the taps weren't turned on correctly ...'

Sound familiar?

Defects result in rework. A defect is something that is not correct. It could be a defective job or product, or incorrect information or service. The defective product, item or information needs to be scrapped, re-done or fixed. Rework is when we have to redo a job or activity because it wasn't done correctly the first time and needs to be fixed. Usually, a defect or the resulting rework is the outcome of a process or procedure not having been followed correctly the first time (or, more commonly, no standard process actually existing in the first place).

HOW MANY TIMES HAVE YOU HAD TO GO AND FIX SOMETHING ON THE FARM BECAUSE IT WASN'T DONE CORRECTLY THE FIRST TIME?

This type of waste is one of the most common wastes on a farm. Every farmer I have talked to—and I am sure if you're a farmer you'll agree—knows that they spend a lot of time in their day dealing with defects, fixing things and reworking. Imagine if everything was done correctly in the first place and we didn't have to spend our days running around re-doing things and fixing mistakes.

Causes of defect waste

There are a variety of reasons why defects, and therefore reworking, happens:

- incapable processes
- no standard work
- equipment malfunction
- poor design
- insufficient training
- inadequate tools or equipment
- poor communication
- poor maintenance
- people error.

A Lean Farm example

Milk down the drain

We have had several instances where milk has ended up down the drain. I'm sure we aren't the only ones. All of these could have been prevented through better standardisation or built-in quality. One instance was the result of taps not having been checked properly and the milk going into the drain instead of the milk Vat.

Cost: $6000

Examples of defects/rework on a dairy farm include:

- milk quality grades
- scrapped milk
- milk down the drain
- scrapped materials such as expired treatments
- motorbike damage due to insufficient oil
- mastitis in cows
- lame cows
- gates left open so cows leave the paddock
- setting up a break on the wrong side of a trough
- dead calves or cows
- high cow empty rates from poorly done mating
- cows put in the wrong paddocks
- an incorrect grazing plan
- cows getting through a fence that wasn't put up correctly
- broken fence reels.

A Lean Farm example

You can't fool those cows

This is a story one farmer told me recently and it's something that has happened to many farmers. Cows were going to be grazing next to a stack of maize. The team was asked to set up a fence around the maize and to make sure it was live so that the cows couldn't get to the maize. The job was not done correctly the first time round and the fence wire was not hot.

Consequently, all the cows broke through the fence and got into the maize stack. The result was that the team had to spend several hours fixing the maize stack and also fixing fences. In addition, this incurred cost for the farm as a result of wasted maize. This is a good example of a defect or rework on a farm.

There are countless examples of defects or rework on farms. Our days are full of them and unfortunately they add a lot of extra time and cost to our business.

Mastitis in cows can be considered a 'defect'

Lean Farm team activity

With your team, can you think of at least 10 more examples of defects or rework on your farm?

Discuss and record them.

Overproduction

Overproduction is when you produce more of something than is necessary. It could be more product, processes, information or services than required. Although this is harder to see on a farm, sometimes the whole dairy industry can be an example of this waste. For example instances where the global supply of milk (in particular milk powder) on the market is greater than the global demand. This is a result of overproduction. Basically, farmers produce as much milk as they can and ship it to their dairy company regardless of whether there is consumer demand for it or not. This can result in too much milk being available on the market, resulting in either it being scrapped (and we have seen countries literally pouring milk down the drain) or driving lower milk prices.

On a more local level, during peak season in the Southern Hemisphere (which is between August and October, just after spring calving) there is a big output of milk, when milk prices are at their lowest. When there is a shortage of milk during autumn (April to June) farmers are paid a premium for their milk. Another example would be too much grass grown and not being able to harvest or graze it properly.

Causes of overproduction

There are many causes of overproduction, among them:

- pushing product out
- poor planning
- inaccurate decision making
- an inability to control the amount produced
- not working to accurate demand/forecast

- poor communication
- poor understanding of customer demand.

Examples of overproduction on a dairy farm include:

- producing too much milk/milk powder when there is no demand
- having too much grass and not being able to use it
- an overproduction of crops
- too much colostrum milk
- too many calves
- too much silage.

Waiting

Waiting is another waste entrenched in all types of farms. In our everyday work we spend *a lot of time waiting*. It is one of our biggest wastes. Would you agree?

Waiting is where people, materials, cows and equipment are not being utilised because they are waiting for something or someone. On the farm we wait for other people, equipment, materials, customers, suppliers, cows or information.

Topping grass due to a surplus

Source: Grant Matthew

Causes of waste resulting from waiting

Time can be wasted if you are waiting because of:

- unbalanced processes
- poor communication
- not having standardised work
- inconsistent work methods
- unclear roles
- unclear expectations
- poor planning
- not having standardised work.

Examples of this type of waste on a dairy farm include waiting:

- for others to show up to a meeting
- for cows to get to the shed
- on information from suppliers about repairs
- on materials such as lime flour to arrive
- for someone to get medications last-minute from the vet
- for the milk vat wash

Richard waiting for the vat wash to finish

- for someone to tell you what needs to be done next
- on someone to be able to do a job.

Generally, as you can see, waiting can also lead to frustration, rushing around last minute to get a job done in less time and with less cost.

Lean Farm examples

No-one likes waiting around

Here are some examples of things that have happened on our farm. Don't forget that farms work with suppliers and customers so these third parties can be affected by our poor farm processes or lack of organisation.

Example 1

We needed to take calves to the sale yards in the morning. Unfortunately, no-one had prepared the calves the day before so they were still in a paddock somewhere far away. This resulted in waiting for someone to bring the calves to the pens so they could be loaded onto the trailer and taken away. Fortunately, we made it to the sale yards just in time!

Example 2

Our lime flour supplier came to deliver lime flour with two trucks. We hadn't cleared any space for it and there were motorbikes and lots of other stuff in the way. The truck driver phoned us, annoyed that he couldn't unload the lime anywhere and we had to go and clear a space. The supplier ended up complaining to us that they had two trucks sitting there waiting for us and it had cost them money in waiting time.

Non-utilised people, intellect and resources

This waste is usually very hard to see, but it is one of our biggest wastes. It consists of two key things:

- not understanding and utilising the skills of team members in the right place
- not discussing or listening to team members about ideas they may have for improvement.

At the end of the day, the people doing the job see all the problems and frustrations every day and are the best source for ideas on how to improve. If we don't take advantage of their knowledge, we miss out on opportunities for improving our business and potentially saving time and money.

Similarly, if we don't understand what skills our team have — and therefore are under-utilising our team members or using them in inappropriate roles — we're not getting the maximum benefit from our team. This will also negatively affect team members' job satisfaction.

Causes of non-utilised people, intellect and resources

The risks of not using people and resources effectively are significant, so ensure that you're not guilty of:

- not engaging with your team
- not listening to people
- not asking people for their ideas
- not empowering your team
- not understanding people's skills
- not doing anything with ideas that are raised.

Examples of non-utilised people, intellect and resources on a dairy farm include:

- a team member who has never driven tractors being responsible for doing feed out
- assuming that a team member knows how to check a paddock and not explaining it properly
- not training new employees correctly or at all
- using a trained mechanic to grub thistles

- using a team meeting to give instructions and not asking for team input
- a farm that has no team meetings at all
- telling people what to do and how to do it instead of asking 'What do you think?'
- being too busy to take time out to talk individually to people and listen to what they have to say (e.g. during a one-on-one catch up)
- not giving a team member with a special interest in pasture the opportunity to be involved in pasture walks
- giving a negative response if someone suggests an idea or raises a problem.

Transportation

Transportation is about moving stuff around—the unnecessary or excessive transportation of materials, products, items, equipment or cows. Our farms are full of this type of waste. Most of the time we're moving stuff around because our farm layout is not optimal so workshops or facilities such as calf pens are far away from each other. This means we have further distances to travel to move calves to calf pens or maize to the feed pad.

Causes of transportation waste

Transportation waste can be a result of:

- long distances to travel
- layout and design
- poor standardisation of work
- ad hoc work
- inadequate workplace organisation
- poor planning
- poor communication
- not thinking about the best location for things (e.g. at point of use).

Examples of transportation waste on a dairy farm include:

- moving calves around to different paddocks
- tractors moving silage from the silage stack to the feed pad
- driving milk cafeterias around to feed calves
- transporting bark chips around to pens
- moving effluent pods around
- moving irrigators around
- moving hay or baleage to paddocks.

Lean Farm team activity

With your team, can you think of at least 10 more examples of transportation waste on your farm?

Discuss and record them.

Ideally, we need to think about how we can improve our layout and reduce unnecessary transport.

Transporting colostrum milk and cafeterias from the dairy shed to the calf pens

A Lean Farm example

Wasted motion

I was sitting in my office one day doing some work at around 5.45 pm, by which time our team should have finished for the day. I looked out the window as one of our team members drove past with a tractor to the hay/calf sheds that are near the house. About 10 minutes later he drove back past the house to the road and in the direction of one of our dairy sheds about 500 metres up the road. Around 15 minutes later he drove back up past the house, then 10 minutes later back down to the road and dairy shed. And, yet again, another 15 minutes later, he was back again. I was wondering why on earth he was driving back and forth so I looked out the window and noticed he was taking the calf bark chips that had been placed near the calf pens by our house one front-end bucket load at a time to the calf sheds that were about a kilometre away next to the dairy shed. This back-and-forth trip happened a total of five times and by the time he had finished it was around 7.30 pm. This is a perfect example of complete waste. If he had had the right equipment, he could have taken the bark in one lot rather than having to make five trips and finished the job in 30 minutes, not 105 minutes. Better still, if we had asked the supplier to dump the bark chips in two loads so that they were at the point of use next to each lot of calf pens, all the transportation between sheds could have been avoided. The team member could have finished work at 6 pm.

Figure 4.7 (overleaf) shows how much unnecessary transport time our team member spent moving bark chips one bucket at a time and having to go back five times.

(continued)

A Lean Farm example *(cont'd)*

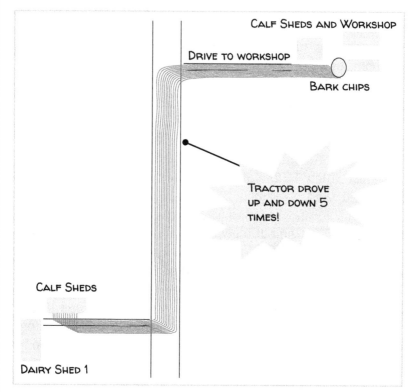

CALF SHEDS AND WORKSHOP

DRIVE TO WORKSHOP

BARK CHIPS

TRACTOR DROVE UP AND DOWN 5 TIMES!

CALF SHEDS

DAIRY SHED 1

Figure 4.7: driving back and forth unnecessarily wastes time

Total wasted distance travelled: approx. 10 kilometres!

Total wasted time: 75 minutes

Inventory

This form of waste is again harder to see on a farm, but it most certainly is there. Inventory is when you have an excess of materials or product that you don't need—in other words, you have more of something than you need to get the job done or than there is demand for. When there's too much milk on the market, many producers end up with stockpiles of milk powder sitting in warehouses. This is inventory. Usually, inventory is the result of producing too much or ordering too much. Inventory wastes space as you need to store stuff that you potentially won't use. Furthermore, many medications and products on the farm have an expiry date, which means if you bought more than you need you may need to throw out expired items. This is obviously money down the drain.

Causes of inventory waste

Inventory waste can be the result of:

- buying too much of something
- producing too much
- no/poor planning
- unclear ordering systems
- inaccurate forecasts of how much is needed
- poor communication
- unclear requirements.

Examples of inventory waste on a dairy farm include:

- too many medications that expire before they are used
- too many inflation spares that end up dirty/damaged
- boxes of saline solution that expire before they are used
- metabolics
- boxes of teat wipes left over from dry cow therapy that end up drying out
- bags of unused grass seed that get wet and mouldy or eaten by birds
- stockpiles of milk powder sitting in warehouses.

Motion

Motion is one of the most common wastes on a farm. Motion is related to human movement and is any movement that does not add any value to the product or customer. Pretty much everything we do on a farm requires some sort of motion. Motion is inherent in farm work. But ...

Would you be happy to pay someone to walk or drive around all day? Would the customer?

Well that's essentially what most farmers are doing. If we followed any of our team members around all day, we would probably see that a big chunk of their day is spent walking or driving around.

Unfortunately, while some of our motion is inherent because of the farm's layout and infrastructure restrictions, most of the motion done on farms does not add value for the customer. In fact, a big portion of the motion that takes place on farms is completely unnecessary and comes about because of poor systems, layouts, work processes or planning.

Causes of motion waste

Motion waste can be the result of:

- a poor work area design
- poor plant/equipment or facility design
- a lack of workplace organisation
- a lack of standard processes
- poor process design

An example of motion—Mat carrying buckets of colostrum milk to feed calves

- large distances to travel across the farm or between farms
- sheds and infrastructure locations/layouts
- grass fed systems—namely, having to move cows
- a lack of technology
- a lack of planning or communication
- tools/materials not being available where needed.

Examples of motion waste on a farm include:

- looking/searching for tools
- having to drive to workshops or maize stacks because of their location
- having several farm blocks or dry stock blocks to drive between
- moving cows and following them down races
- having to stretch to reach Vat taps
- hosing down yards
- walking around platforms to fix cups that are sucking air
- putting up breaks
- forgetting a tool you need and going back to get it
- going to a paddock twice to do two different jobs that could have been done at the same time
- three people going to do three different jobs on the same part of the farm, instead of one person doing all three jobs.

Lean Farm examples

Have your tools on hand

Example 1

We have a dry stock block that is about 20 minutes away. One day, Mat had to shift some stock and he drove over to do this. While there, he noticed a running trough.

(continued)

A Lean Farm example *(cont'd)*

Despite having an old woolshed and quad on site, he had no tools with him or on the dry stock block with which to fix the trough. So, he had to drive all the way back to the dairy farm, get the tools he needed and then drive back to fix the trough. This consumed an additional hour of driving time. In the end, what could have been a 10-minute job became a 1 hour and 30 minute job. This is a classic example that happens to many farmers and could be solved simply by having the right tools in the right place—particularly tools needed for common farm maintenance problems.

Example 2

Going back to my defect example of several team members having to go and fix fences that were damaged because the cows got into the maize stack, the story gets better...

Six guys arrived at the paddock to fix the fence, and then realised that between them they only had one hammer! So the guys had to go back, find more hammers and return to the paddock. All *wasted* motion!

By having clear standardised processes, more planning, better team job allocation and communication, the right tools in the right place and carefully designed facilities and farm layouts, we can cut out significant motion from our day. This will give us time to focus on more important activities.

Lean Farm team activity

With your team, can you think of at least 10 more examples of motion waste on your farm?

Discuss and record them.

Extra-processing

Extra-processing is a bit harder to see on farms, but it does exist. This is where a process or person works hard, but not necessarily smart, or does more than is necessary to deliver the expected outcomes—in other words, when we add processing steps to our work that are unnecessary and do not add any value for the customer because no value is added to the product/ service. This creates a waste that's not always easy to see, and can often be mistaken as part of the process.

Causes of extra-processing

Extra-processing can be caused by:

- unclear customer/business requirements
- poor processes
- untrained staff
- unnecessary process steps
- overcomplicated processes
- a lack of standardisation
- excessive information
- new technology being used inappropriately or insufficiently.

Examples of extra-processing on a dairy farm include:

- recording too much information that is never used
- duplicating information
- a very high cow condition score
- an extremely low milk cell count (e.g. at 10 000 when 150 000 is perfectly acceptable)
- too many people having to approve something
- approval being required unnecessarily
- reports with too much information that is generated but not used.

A Lean Farm example

Reducing motion waste

After learning about the 8 wastes, one farmer observed a team member filling the silage wagon. He noticed the amount of driving back and forth between various piles of feed this involved. This was not only wasting diesel but also a lot of time. He made changes to the layout and bought a larger bucket for $1000 to reduce the amount of movement.

This saved him more than $20000 a year in labour and diesel costs.

Saving: $20000 per year

Finding waste

Now that you know what 'the 8 wastes' are, you are probably thinking this is just plain old common sense. Nothing new. Well unfortunately while it looks simple and sounds simple—and you're right, it is common sense—most of us can't do it very well. This is why our farms are still full of so much waste. If it were so simple, businesses wouldn't still have 95 per cent or more waste in their processes. The difficulty is in firstly *seeing* the waste around you and secondly *accepting* that there is waste.

Waste is intertwined with everything we do. It becomes a part of our everyday process and we don't even notice it. Therefore it becomes very difficult to pick out what is waste and what is value.

LEARN TO SEE WASTE.

Because we're so used to doing things the way we do them—'we have always done it this way'—we think that's the way they have to be done.

This obscures our vision and we don't scrutinise and question the processes around us enough to identify whether things truly need to be done or whether they are in fact waste.

Lean tools and thinking help us to recognise waste in all its forms and eliminate it from work activities.

Lean Farm team activity

A 'waste' walk

With your team, take a walk on your farm and follow these steps. Use the waste walk template example (overleaf) and ask each person to record what they see.

1. 'Go, look, see' the farm.

2. Look at different parts of the farm, people, vehicles and animals.

3. Observe people at work, animal/tractor/equipment/vehicle movement.

4. Identify as many of the 8 wastes as possible.

5. Identify any overburden or unevenness.

6. Record the wastes on your sheet along with a description of what you saw and where.

7. Discuss and identify what could be improved to eliminate the wastes you observed.

(continued)

Lean Farm team activity *(cont'd)*

FARM WASTE WALK

WASTE	WHAT DO YOU SEE?	HOW TO IMPROVE?
DEFECTS (REWORK)	LOTS OF LAME COWS	CHECK ALL RACES/ BRAINSTORM SESSION
OVER-PRODUCTION		
WAITING	PERSON 1 IS WAITING FOR COWS TO ARRIVE ON YARD	LOOK AT WORK BALANCE BETWEEN PERSON 1 AND PERSON 2
NOT USING RESOURCE	WE DON'T TALK ABOUT IDEAS OFTEN	PUT IDEAS ON AGENDA FOR TEAM MEETINGS
TRANSPORT	HAVING TO MOVE MINERAL PALLET FROM PADDOCK 51 TO SHED	BUILD SHELTER FOR MINERAL PALLET NEXT TO MAIZE STACK NEAR SHED
INVENTORY	LOTS OF EXPIRED MAG BAGS	CHECK USAGE VERSUS ORDER QUANTITIES
MOTION	PERSON 2 IS BENDING TO REACH PAINT CANS ON GROUND WHILE MILKING	PUT PAINT CANS ON POLE AT RIGHT HEIGHT
EXTRA-PROCESSING		

A team member filling the silage wagon—this involves a lot of movement back and forth

Stand still and observe

A while back when I was working at Toyota I learned a very valuable technique to help 'see' waste around you. In Japanese, it was called *tachinbo*. Essentially, it means finding a central spot on the farm from which you have a good view, drawing an imaginary circle around your feet and standing still in that circle observing the environment around you.

STAND *STILL* IN A CIRCLE LIKE A STATUE AND *WATCH.*

The reason why you stand in one spot is that when you move around, you miss things. By standing in one spot for a period of time you will see the interactions of machines, people, materials, animals and so on. Of course, if you are a one-person farmer, this probably isn't such a valuable activity...although it could be a good opportunity for self-reflection!

While doing this activity, *don't* talk to people or ask them questions—just watch silently and see what happens. Then ask 'Why?' Always *challenge* what you do! What can we do to improve?

Some things to look out for when you're doing this activity are:

- *movement.* Look at the movement of people, animals, materials, machines, tractors and other vehicles—is it smooth? Are people racing around back and forth? Are they bending, overburdening, trying to fit into difficult spaces?
- *standard work.* Are things being done the same way consistently or is everyone doing things their own way?
- *the 8 wastes.* Look out generally for movement, transport, waiting and so on.
- *communication.* How are people communicating with each other? How do they know what needs to be done, when and who will do it?
- *interaction.* How are people, animals and equipment interacting with each other? It is chaotic or smooth?

79

- *facial expressions.* Take note of people's faces and expressions. This will often tell you a lot about how they are feeling, such as whether they are frustrated, confused or tired.

Lean Farm team activity

Tachinbo time

Ask each team member to find a spot on the farm and spend at least 30 minutes standing *still* like a statue.

Ask them to record what they see, including any problems, waste and opportunities for improvement.

As a group, discuss the observations made by each team member. Identify any potential actions that could be taken to improve the situation.

Value-added vs non–value added words

One thing that can help you identify whether an activity is value-added or non–value added (in other words, *waste*) is to describe the activity or task, starting with the action verb. As an example, if I am watching a team member do some cleaning in the shed I might observe one task as being 'walk to the Vat', where 'walk' is the action verb. The type of action verb that you use tends to give away whether that particular task is value-added or non–value added (waste). Figure 4.8 gives examples of each.

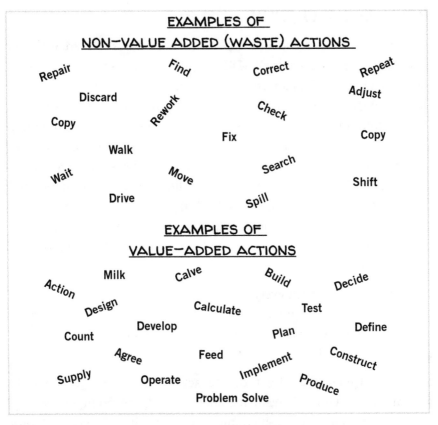

Figure 4.8: value-added vs non–value added words

Example: Mat feeding the calves—'feed' is a value-added verb

Day In the Life Of (DILO)...

The final useful technique for identifying waste I'd like to share with you is called the 'Day In the Life Of' (DILO). (Although, it is really just a fancy name for a time and motion study.) I came across this activity when I was doing some work for a client a few years ago. Basically, it means putting yourself into someone else's shoes by observing and recording everything they do over a period of time. The idea is that someone finds another colleague at any random point in the day and follows them around for at least 30 minutes, observing and recording every activity performed during that time in time intervals of, ideally, one minute. By doing this you can capture *every* single activity that is done including walking, waiting, moving, driving, thinking and doing. It becomes a snapshot of a typical day.

Once you have recorded every task/action and the time it took over the observed period, you can identify which tasks are value added and which are waste. From there you can easily calculate the total percentage of waste in the observed period.

I can almost guarantee, if you do this activity accurately and honestly several times, with any team member, and at any point in the day, you will recognise a very common pattern. It will become very apparent that no matter what is happening, and who is doing what and when, there is a significant amount of time spent on waste *every day* by *everyone*.

A Lean Farm example

Don't waste your time

One day I had some spare time between a meeting and kids waking up so I jumped into the ute and headed for one of the dairy sheds to do a DILO. I bumped into our manager, and asked if it would be okay for me to just follow him around for an hour and a half or so. Here is part of the DILO (with one-minute intervals) that I filled out while I was following him around.

FARM DAY IN THE LIFE OF (DILO)																															
DATE: 12/03/18				FOLLOWING: MANAGER					START TIME: 11:04 AM					FINISH TIME: 12:04 PM																	
ACTIVITY/TASK	1	2	3	4	5	6	7	8	9	10	11	12	13	14	15	16	17	18	19	20	21	22	23	24	25	26	27	28	29	30	..	99
Put stuff in locker																																
Drive home																																
Get ready																																
Drive to town																																
Wait for technician																																
Fix Reader																																
Tech explaining prob																																
............etc.																																

From the DILO I could work out all the activities that were non-value added as well as the value-added ones. Most of the one-and-a-half hours was spent driving to a supplier to fix a pasture meter reader that a team member had broken by driving too fast and running into wiring. This was an example of a defect resulting from not taking care and wasted at least 40 minutes of a manager's day. Other than changing behaviour, potentially a 'poka yoke' could have been used to prevent this defect. We learn about this Lean tool in chapter 10.

All up, out of the 99 minutes that I followed the manager, 47 minutes were spent on waste activities: fixing a defect, moving, waiting and transport. This represents a minimum of 47 per cent waste in only one and a half hours. Imagine if I had followed him the whole day?

Steps to eliminate waste

As I mentioned earlier, it isn't as easy as it seems to eliminate waste from your farm. It is easy to be influenced by old habits or workplace conditioning. Importantly, you and your team need to be open to the idea that waste is all around us and accept that we aren't perfect and there are plenty of inefficiencies in what we do.

Figure 4.9 outlines the steps that will help you do this more successfully.

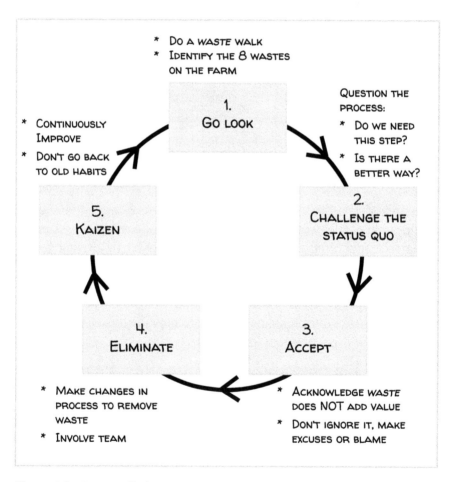

Figure 4.9: steps to eliminate waste

Lean Farm team activity

Waste challenge

With your team, do the following:

1. Identify any farm activity that needs to be completed today.

2. Video 5–10 minutes' worth of this activity being performed.

3. Watch the video with the team (replay and stop it as many times as needed).

4. Identify as many of the 8 wastes as possible in the process.

Then answer the following questions with your team:

1. What percentage of time is spent on waste in the activity you observed? Be *honest*!

2. Calculate the actual percentage of waste by pausing the video to capture time intervals of elements of waste versus value-added activities.

3. What can be done to improve this process?

Benefits of eliminating waste

There are countless benefits to eliminating waste and I am sure you can appreciate them without me listing each one. A few of these are listed in figure 4.10 (overleaf). Ultimately, it will give you and your team more time, with less effort, and make farming less frustrating and more rewarding. It will also, of course, result in a more productive, profitable and sustainable farm.

Benefits of eliminating *waste*

Customer benefits

* Product that meets all expectations
* Satisfied end consumer
* Competitive product

Business benefits

* Reduced costs
* Improved quality
* Improved productivity
* Increased flexibility in operations
* Improved reputation

People benefits

* Easier work
* Lower work hours
* Safer work environment
* Consistent workload
* Fewer frustrations and chaos
* Increased pride in work area and work quality

Animal benefits

* Less walking
* Less stress
* Better health
* More time in paddocks eating

Figure 4.10: some of the benefits of eliminating waste

Summary

To truly critically analyse what we do and try to eliminate as much waste as possible we need to really value our time and money. In fact, let me ask you this:

How valuable is your time and money to you?

How you answer this question will determine how hard you will work to squeeze out as much waste as possible from your processes. If you are happy to spend an extra 20 minutes filling the silage wagon each day and you don't care that you could be spending that 20 minutes on more value-adding activities or with your family, then you won't be so determined to eliminate waste. If you would rather be at home with the kids, you will want to remove every minute of non–value added work.

So it's really up to you. Why don't you set yourself a target of eliminating at least 10 per cent of the time spent on every activity you do on your farm to start with? I'm certain that with no additional technology, resources or spending, you can do this by simply analysing, challenging and cleverly improving the process steps through eliminating waste.

It's not a question of 'this can't be done' but rather a question of 'How can we do this?' because there is *always* a way if you want to find it.

Make sure you time your particular activity/process before improvement and after improvement so that you can measure the real tangible benefit eliminating waste will have on your farm.

Lean Farm team quiz:
THE 8 WASTES

1. What do VA and NVA stand for?

2. What are the eight sources of waste?

3. What are non–value added activities also known as?

4. Who benefits from waste elimination?

5. What do the 3 Ms stand for?

6. How would you define a VA activity?

7. Why is 'non-utilised people' waste so important?

8. Why is waste so hard to see?

9. What is incidental work?

10. What three verbs would suggest a VA activity?

Lean Farm action plan:
THE 8 WASTES

These actions are included to provide some simple guidance for eliminating waste on your farm. They are aimed at giving you a little bit of motivation and direction if you need it. Of course, the idea is for you to do as much as you can to gain the maximum benefit of waste elimination on your farm.

1. Introduce 'the 8 wastes' to your team.

2. Go on a 'waste' walk.

3. Look at different parts of the farm: people, vehicles, animals.

4. Do a *tachinbo* activity.

5. Take a video of at least two processes and analyse them with the team to identify any of the 8 wastes.

6. Do at least two DILOs on two different people for at least one hour each. Record the results.

7. Calculate the value-added work in each case and discuss any improvements.

8. Identify as many of the 8 wastes as possible in your processes, including overburden and unevenness.

9. Discuss and agree to ways of eliminating this waste with your team.

10. Action: create an action plan and make necessary improvements.

11. Take before and after photos of the waste and changes.

Hopefully this overview of how to identify and eliminate waste has armed you with the tools you need to address waste on your farm.

In chapter 5 we look at the 5S methodology — the second of my 10 Lean tools for transforming your farm.

Chapter 5

5S workplace organisation

This chapter is probably one of the most commonsense chapters in the book. The tool that I introduce is something that should be the norm for every farmer and business. While it is so simple and can make every person's life much easier and more productive, it relies on human behaviour and habits to change and this can make 5S hard to implement or sustain.

How 5S came about

Have you ever started a job only to realise you don't have all the tools you need to do it? So you waste time running around trying to find what you need. I am guessing your answer is 'yes' and possibly even something along the lines of 'every bloody day'. Well that's how it was on our farm (and still is in many areas as we slowly introduce 5S) and I know this happens on just about every other farm — and of course outside farming too.

There is nothing more frustrating than when you urgently have to tighten a pipe and you just can't find the right wrench. You get on the phone and call the team asking if anyone has seen it and perhaps 30 minutes later you might be lucky enough to locate it. Well that was a waste of half an hour, right? Wouldn't you have rather spent that time doing something else?

The 5S methodology has been used for decades at Toyota to support their just-in-time (JIT) system by ensuring that the right tool is in the right place, and people aren't wasting time looking for things. (We discuss

JIT in detail in chapter 12.) It is considered important 'house-keeping' that creates a clean and efficient work environment. (Interestingly, it is widely believed that Henry Ford developed the precursor to 5S with his CANDO approach.)

Since being introduced formally to the world in Takashi Osada's book *The 5S's* in 1991, 5S has become one of the building blocks of Lean and is one of the most effective Lean tools for improving Lean businesses' productivity and waste elimination. Not only that, but anyone can do it—and it is low cost and easy. Before we delve into the meaning of 5S, try this number game with your team.

Lean Farm team activity

The 5S number game: Round 1

Give each team member a printout of the number chart below. You will also need a timer and a flipchart (or similar).

1. Numbers 1 to 80 are scattered on the chart.

2. Ask the team to find and cross out numbers 1 to 50 in sequential order (i.e. 1, 2, 3, 4 ...).

3. Tell them that the target is to do this in three minutes.

4. Start the timer.

5. Record, on a flipchart, the time taken for the first person to finish crossing out all 50 numbers (if anyone finishes within three minutes).

6. Record the time when most other people finish counting.

7. Plot the first time on a graph on the flipchart and call it 'Round 1'. Stick it to a wall.

Now discuss these questions with your team:

1. Why could most people not finish in under three minutes?

2. What was the problem?

3. What could have made this activity easier?

What is 5S?

5S is a process for organising, cleaning and maintaining a workplace.

The aim of 5S is to create a safe, standard, efficient and effective workplace for all employees. It is *not* just about tidying, washing the yards and floors or making the farm look pretty.

5S should form the foundation of your approach to the job you are employed to do. It is a fundamental part of a Lean Farm and helps to identify and eliminate waste.

5S will help you to stop wasting time, cut costs, be more productive and run a better farming operation. To be successful and effective, 5S needs to become a standard way of working on your farm and *everyone* must be on board and committed to sustaining 5S.

5S IS *NOT* A 'ONE-OFF' ACTIVITY; IT SHOULD BE PART OF YOUR EVERYDAY LIFE ON THE FARM!

5S was originally only 4S at Toyota, and represented the Japanese words Seiri, Seiton, Seiso and Seiketsu. It eventually developed into 5S with Shitsuke added. These have been roughly translated into five English words also starting with S: Sort, Set (in Order), Shine, Standardise and Sustain (see figure 5.1). (There are several versions of these translations floating around but I think these five words capture the essence well and resonate with people.)

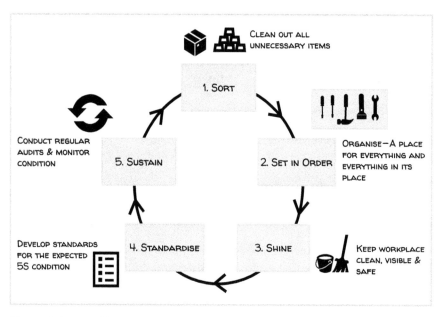

Figure 5.1: the 5S steps

5S can be very subjective because everyone has a different idea of what a well-organised workplace looks like. This is why the 5 steps are important—they will help you define a clear standard around what 'good' means on your farm. They will help every team member stick to the same standard so that your workplace remains at a consistent level of organisation and cleanliness no matter who is at work.

A Lean Farm example

The missing calving pulley in the night

One night in the middle of winter during calving, Mat went out to check the cows. I went to sleep at about 10 pm. At 1 am I woke up and Mat still wasn't home. Of course, I was worried and phoned his mobile several times but there was no answer. I waited another hour and still there was no sign of Mat. So I got out of bed, put on my dressing gown and crocs and drove over to the shed to look for him. I found him out in the wind and rain trying to help a cow calve.

He asked me to go and find the calving pulley that was apparently somewhere near the platform at the other shed. So I drove around to the other shed—there I was, running around a shed at 2 am, in the rain and wind, six months pregnant, in my dressing gown like a mad person looking for some pulley (and I didn't even know what that was). Of course it couldn't be found just when you needed it! I remember thinking, 'This is just ridiculous; how on earth isn't there a standard place for this thing so it can be easily found when you need it?' In the end, I couldn't find the pulley and Mat had to improvise using some ropes around a fence rail to get the calf out, but obviously it took him a lot longer than it would have if he had had the right tool on hand.

Where to use 5S

There are many places on the farm where 5S should be applied to make your team more productive:

- workshops
- pump rooms
- staff rooms
- store rooms
- chemical rooms/ cupboards
- yards
- maintenance areas
- dairy shed
- paddocks
- races
- team housing and gardens
- farm vehicles
- silage stack area
- farm offices
- irrigator sheds.

Lean Farm team activity

A 5S hunt

1. Ask your team to split up into groups of three to four (if your team is big; otherwise just have one group). Explain that each group must work together and can't separate. They have 10 minutes to find the following items on your farm:

   — 13 mm pipe wrench

   — red spray can

   — mastitis drug

   — calving chains

   — thistle grubber

   — inflation spare

   — sockets

   — cow tail trimmers

   — hoof scraper

   — hammer.

2. Start a timer.

3. Teams should take a photo of each item and record where it was but can leave the items there. They must be back within 10 minutes.

4. When they return, discuss with the teams what items they found. How easy was it to find the items? What could have made it easier to find these items? Were the items in their correct place?

Benefits of 5S

There are substantial benefits to applying 5S (see figure 5.2 for an example). I can honestly say that this Lean tool alone can make a tremendous difference to the efficiency of your farm and can save hours in your team's day. Most of the farmers who have completed my LeanFarm training program absolutely loved this tool and said it saved them a significant amount of time each day. Specific benefits include:

- it eliminates waiting time — it is harder for things to go missing as there is a place for everything and everything is in its place
- it eliminates or reduces waste within our farm processes and tasks
- it reduces variation because it is easier to distinguish what is or isn't normal
- it improves the ability to manage by making the workplace more visual and clean because any deviations are identified immediately
- it improves quality and reduces errors/rework
- it means less time is spent searching for things
- it improves process flow
- it makes jobs easier to do
- it improves safety
- working in a clean environment lowers stress levels because people can find things
- it is easier to train new staff

Figure 5.2: one of our workshops before 5S. Does yours look like this?

- it eliminates rubbish and clutter because there is nothing extra/unnecessary lying around
- it provides a more enjoyable and pleasant work environment for everyone
- it simulates pride in the workplace.

I have worked with clients in all sorts of industries and 5S has even saved lives in industries such as healthcare (and probably farming too).

Are you spending time looking for things?

Are things constantly going missing, *costing* you money?

Do you have what you need when you need it?

Lean Farm team activity

Identifying 5S on your farm

Ask your team to walk around different areas of the farm. This could be the shed, workshops, yards, silage stack area, feed pad, paddocks or offices. As you are walking around, make the following observations:

— How tidy and well organised is your farm currently?

— Rate the farm from 1 to 5 in each area (shed, office, workshop, feed pad, yards, paddocks etc.) with 1 being terrible (disorganised and messy) and 5 being excellent (clean, tidy and well organised).

— What are examples of good organisation on your farm? (Take photos.)

— What are examples of opportunities for improvement on your farm? (Take photos.)

— How easy is it to find tools and equipment on your farm?

— Do tools, equipment and other items have 'homes'—clear locations where they are always kept?

— Do you have to often run around searching for things? Which things?

— Re-group with your team and share with each other what everyone has observed.

— Record any actions/areas needing improvement.

Let's now take a look at each of the 5S steps in detail to see the phenomenal difference they can make on your farm and how to practically implement each step.

The 1st S: Sort

Often, there are things on our farms that have been there for years and we keep them 'just in case'!

Sorting is about reviewing *all* items that are lying around in the shed, workshop, paddock and yards and asking ourselves:

- Do we need them?
- What are they used for?
- How often will we use or need them?
- When did we last use them?
- Are they still relevant (in date, latest model, useful)?

We then get rid of everything we don't need. *Everything*. We can scrap it or give it away to a new home. This 'S' eliminates hoarding behaviours, which many farmers are prone to.

Why and what?

The idea behind sorting your farm is for your work areas to only contain what's required to do the specific job or process performed in this area. Everything else should not be in the area. The more 'stuff' you have cluttering a workplace, the less efficient you and your team are. In other words, having non-essential items in the workplace will reduce your farm's productivity. It will make it even more difficult to find what you really need among all the junk; you will be storing things that don't work and potentially trying to fix machines/equipment with malfunctioning components; you will be wasting money ordering tools that you already have hidden under piles; and, importantly, your workplace will feel messy, stressful and chaotic. It will not be a professional-looking farm.

To Sort your farm means to look at and sort through each area of your farm and remove any non-essential items including equipment, materials, supplies, tools, machines and general rubbish.

A Lean Farm example

Costly waste

During one of our 5S activities we discovered containers, bottles and boxes full of expired product, including cow treatments. All this product had to be discarded. We hadn't used it because it was lost among the mess and we'd forgotten about it.

Cost: more than $2000

How to sort

Here are five steps that will help you and your team sort your farm.

1. Identify an area of your farm that needs to be sorted. (Work on one small area at a time.)

2. Define the boundary of each area clearly — for example, the workshop or the pump room. (If the area is too big, the job will become overwhelming.)

3. Get the right people involved: those who work in the area or are involved in the area. (Never do this on your own; engage your team.)

4. Clear out absolutely everything movable from that specific area (essential and non-essential things) and place it all in a 'Sort' holding area, which should also be clearly defined and marked out. This means also removing all the items from your filing cabinets, cupboards, drawers and other storage locations, which are horrendous culprits of hoarding.

5. Sort out the required items from the unwanted items. Sort through everything in the 'Sort' holding area and separate it into four piles: 'Needed items (in area)', 'Needed items (in other area)', 'Unsure', and 'Scrap area (to get rid of)' (see figure 5.3). Ensure that the items you are keeping are clean and in good working order, ready to be returned to the relevant work area.

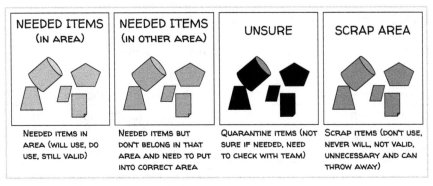

Figure 5.3: sorting items into four piles

When sorting, don't think, 'we might need this someday', but rather 'does the current operation actually need this?' Keep in mind that people become attached to their 'stuff', useful or not, and are reluctant to see the items as waste. And farm owners tend to be the biggest culprits! These people will see the action of throwing items out as waste. Out of courtesy, we should always let someone know what the item will be replaced with, or explain why it must be disposed of.

Things to look out for on your farm:

- defective or excess items that create clutter
- broken pieces of equipment
- old rags and cleaning supplies
- equipment waiting for repair
- outdated posters, signs and documents
- old bottles that aren't labelled filled with unknown substances, which might be oil and turpentine
- dried-up paint cans
- broken fence wires or handles that can't be repaired
- rusty old wires and steel scraps lying around on paddocks
- bits of old pipes and hosing that are cracked and unusable.

It can often be tricky to decide whether you do need something or not. Even more difficult is deciding where something you do need should be located. See table 5.1 for help with your 5S sorting activity.

Table 5.1: deciding what is needed and where to put it

How often is it used?	What to do with it
Never	Give away, sell or throw away
Once or twice a year	Store in a distant place
Once a month	Store in the facility, but away from the 'point of use'
Once a week	Store in the work area
Once a day or more	Store as close as possible to the 'point of use'

A Lean Farm example

Sorting on our farm

We are slowly doing 5S across our farm. One of the first areas we sorted is the pump room at one of our dairy sheds. The whole team was involved in the activity and they essentially drove it so that they had ownership of the process. Figure 5.4 shows some photos of the sorting stage.

Figure 5.4: sorting a pump room on our farm

Surprise, surprise

When you start to do the 5S activity you will find a lot of surprises. You will no doubt discover items that you needed at some point but thought you had long lost. Perhaps, like on our farm, you will discover things you thought were lost and that you bought again only to discover the original item under a pile of old hoses during your 5S.

You will most likely also find some items that you had paid for but are now expired or rusted and unusable. A bit of wasted money, right? This is exactly what happened on our farm too. As you can see in figure 5.5, in the pump room alone we discovered bottles and boxes of magnesium, copper and various other substances that not only should not have been there but also were long expired. The cost for us: more than $2000 wasted.

Figure 5.5: finding expired products worth more than $2000 during 5S on our farm

Don't forget the electronic stuff!

Physical things might be easy to see and sort. But these days our electronic world is ever more important. Much of our information and data—such as production, processes, quality, animal health, costs, invoices, HR, safety and maintenance — is stored electronically. Therefore, it is just as critical to have a very well-organised electronic system (as shown in figure 5.6).

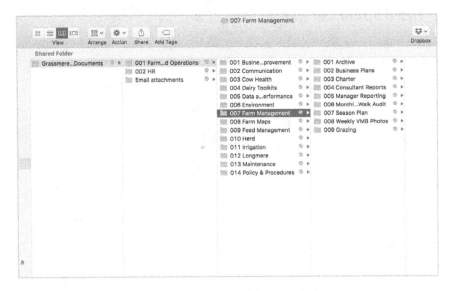

Figure 5.6: make sure you 5S your hard drive regularly

We need to make sure we regularly 5S all our computers and hard drive data, information and emails so that we can easily find what we need when we need it. Think about your computer or electronic files:

- How well organised are your electronic files?
- Do you know where everything is stored?
- How easily can you find what you need?
- How easily could one of your team find a document they need?
- Is all your electronic information actually needed and used regularly?
- Are files or documents duplicated in multiple locations and folders?
- Do you sort and delete emails?
- Are folders accurately labelled and dated?
- Is there version control on all documents?

Lean Farm team activity

The 5S number game: Round 2

We have now done some sorting, so let's play the number game again. Give each team member a printout of the following.

NUMBERS 1 - 25	NUMBERS 25 - 50

7 1 18 5 13
9
15 23 11 20
24
19 14 6 2
12 22
17 10 16 8
25 4 27 3 21

40 28 43
45 31 37 47
26 39
41 35 44 30 34
46
36 29 27 38
50 32
33 48
49 42

1. All the numbers that are not needed have now been scrapped. We have also sorted the remaining numbers into two piles: numbers 1–25 and numbers 26–50.

2. Ask each team member to find and cross out numbers 1 to 50 in sequential order (i.e. 1, 2, 3, 4...).

3. Start the timer.

4. Record on a flipchart the time taken for the first person to have all numbers crossed out.

5. Record the time when most other people finish counting.

6. Plot the first time on the same graph as previously on the flipchart paper and call it 'Round 2'. Keep it displayed on the wall.

Now discuss these questions with your team:

1. From the graph, can you see that on average it took about half the time that it did at round 1 to sort the numbers?

2. Why was it so much quicker?

3. Did sorting help?

4. Can you relate this to your farm workshop: if half the stuff wasn't in there, how much easier would it be to find something?

The 5S number game is a simple activity that demonstrates just how effective the first 5S step can be towards improving your farm productivity by helping your team save significant time each day. While it may be just a game, it is very reflective of our real work environment. Now that you have an appreciation of the impact the first 'S' can have, it's time for you to try it out for real on your farm.

Lean Farm team activity

Sorting your farm

1. Agree on an area to sort with your team (e.g. workshop or cups-on area).

2. Go to the area.

3. Agree on and mark out the work area boundary.

4. *Take before photos* (this is important so that you remember the original condition and can create a standard and also celebrate the hard work).

(continued)

Lean Farm team activity *(cont'd)*

5. Make sure all equipment is isolated (we don't want any safety incidents).

6. Display notices on power switches if required.

7. Agree and mark out a 'Sort' holding area—an area where all the items you remove will be placed until sorted so that they are not in the way of your farm work and don't create any hazards.

8. Remove *everything* (that's not fixed to the ground) from the workplace area and place it in the holding area.

9. Decide which items are essential and which are not required.

10. Sort the items into four piles as shown in figure 5.3.

11. Take action on each of the four piles of items:

   — Talk to the relevant people about the items in the 'Unsure' pile and allocate accordingly.

   — Move the 'Needed items in other area' pile to the other areas.

   — Dispose of or give away the 'Scrap' items (take them away immediately before people start collecting things 'just in case' out of the pile).

   — Keep only the 'Needed' pile in the appropriate areas.

Before and after sorting the cups-off area at our Bodmin shed

The next 'S' in our 5S cycle is Set in Order. However, before we learn about Set in Order I am going to skip straight to the third 'S', Shine. This isn't to confuse you, but rather it makes more sense practically for your implementation.

The 3rd S: Shine

I have deliberately skipped the second 'S'—Simplify—to talk about Shine first because it is more logical and practical to do Shine before Set in Order. We want to ensure that our work area is spotless before bringing things back into the area.

Shine is about making everything 'spic and span' (Toyota's words for 'shine'). It is about ensuring that the workplace is clean and in working order.

What and why?

Shine is more than just cleaning. Importantly, it is vital maintenance. It is also a form of inspection of the workplace that examines the current condition of equipment and facilities in the work area and identifies whether there are any problems.

While performing Shine you will identify things that are broken, such as cables, sockets, wiring, lights, switches and meters. You will also identify the sources of dirt, contamination, oil leaks and water and milk leaks so that they can be fixed and a clean and tidy workplace can be maintained.

If equipment such as filters, coolers, pumps, wiring, hoses and materials are full of dust, dirt, milk residue, spider webs, bird poo or grease, or are broken, then the equipment functionality, safety, efficiency and life span can be negatively affected. Furthermore, this can also have an adverse effect on your production and quality results. Neglecting a problem or condition can also make it much more expensive to fix once you finally identify it. Doing Shine regularly makes it much easier to spot problems such as water, oil or air leaks or blocked filters as quickly as possible and fix them promptly.

Shine ensures that everything in a work area is clean, functional and ready to be used in an optimal manner.

Shine is fundamental for your farm and has many benefits:

- There is less chance of affecting product quality by contamination.
- It identifies abnormal conditions, problems, defects and maintenance issues.
- Dirty or poorly cleaned machines or equipment tend to break down more often. Unplanned equipment breakdowns result in lost time, poor quality and unnecessary cost.
- Safety is improved—less oil, water, residue and dirt on the floor reduces the risk of slips, trips and falls.
- It demonstrates control of your work area.
- It makes your farm a more professional place to work.
- It improves team morale.
- It shows you have pride in your workplace.

White factories

When you think of a factory, you probably think dark floors, dark walls, dirty green machines, noise, grease, dirt and so on—right? And you would be correct. I've been in plenty of black, dark, dirty factories around the world. However, the factories of today are changing this stereotype. Many new factories—whether for car parts, plastics or food—are being constructed with white floors, white machines and white walls, like the one in figure 5.7. Why? It's not so that the place looks white, but because white shows up any form of dirt, oil leaks, grease, water and so on immediately.

Figure 5.7: a white car factory
Source: © BELL KA PANG/Shutterstock

These are not hidden on a dark floor in a dark corner. By being white, these 'problems' can be immediately spotted and fixed, improving the factory's performance.

DAIRY FARMS ARE DIRTIER THAN FACTORIES.

CHALLENGE: CREATE A WHITE FARM.

Considering that our dairy farms are producing a product for human consumption, it's surprising that many of the farms I have seen are grubbier than factories I have been in. Is this acceptable? Everyone probably has a different viewpoint on this, but what I do know is that having a clean, white farm would be hugely beneficial to you, your people, your animals and your business.

Can you imagine having a white farm where floors, races and platforms are all white? How easily and quickly could you see the stones, dirt, bird poo and oil leaks? If you can't see them because your floors are black or grey, then it's a good excuse to ignore them, right? But if they are obvious and can be spotted from a mile away, you are more likely to do something about them. Probably the only thing you couldn't see easily on white floors, unfortunately, is milk!

How?

Once you have removed all movable items from the area during Sort you can begin cleaning it thoroughly. Here's a checklist you can use.

- Clean everything: floors, pipes, fixed machines and equipment. Clean from the top down, starting with ceilings, and the tops of cupboards and equipment. Large items such as vats, water tanks and pumps should be cleaned from the top down also.
- Clean all the items in the 'needed' pile of the sort area. These will go back into the area during the 'Set in Order' step and should be clean.
- Identify any problems with a red tag and record them in a red tag log (see the section on red tags on page 115).
- Try to fix the sources of any dirt and problems found during the Shine activity where possible.
- Introduce a 'clear desk' policy in offices and employee rooms to get people into the habit of 5S.
- 'Shine' every day: create a standard procedure and frequency for Shine.
- Ensure Shine is done by team members to promote ownership.
- Develop a plan, if needed, by assigning tasks and timing to team members.
- Make Shine a daily habit and part of the culture.

Ask the following questions as you do this activity:

- Are floors, walls, stairs and surfaces clear of dust, dirt and rubbish?
- Is equipment in the work area free of dust, dirt and rubbish?
- Are benches, tables and cupboards clean?
- Are labels, signs, standards and so on clean, accurate and in good condition?

- Are all pipes, cables, wires, control panels and hoses clean and free of dirt, dust, spider webs and milk deposits?
- Are all hidden corners and spaces clean?
- Are cleaning materials and equipment easily accessible to everyone?

A Lean Farm example

Pesky nests

One day a team member noticed that the in-shed feeding system in one of the dairy sheds wasn't working properly. The hopper was not releasing food into each bail so cows were not being fed the required quantity of minerals and maize. This had been going on for a few days, and had resulted in sub-optimal feed for the cows, and hence reduced milk production. After some investigation, it turned out that the feed hopper was full of birds' nests, which were blocking it. If this had been a part of a 5S or Shine checklist, it would have been identified immediately, preventing inaccurate feeding and milk production loss.

Shine every day

Maintaining a clean work area makes identifying problems that were potentially hidden behind rubbish or grease much easier. It also ensures a pleasant, safe and efficient work environment. However, to sustain an excellent Shine condition your team should establish clear guidelines for cleanliness and orderliness. Furthermore, Shine must be incorporated into the daily work routine to ensure that it happens *every day* so that work areas are always in a presentable and acceptable condition.

The 5-minute Shine

Carrying out a '5-minute Shine' involves five minutes of dedicated cleaning or 'shine' activity that is done every day as part of a standard process.

It could be the last five minutes of each shift—or the first five minutes. Checklists detailing the different areas or items to clean are useful so you don't waste time wondering what to do next. You can also develop a schedule to allocate who does what 'Shine' job, where and when, and rotate responsibilities. See the example in table 5.2.

Table 5.2: a 'Shine' schedule

	Monday	Tuesday	Wednesday	Thursday	Friday	Saturday
GARY	SHED	YARDS	SHED	WORKSHOP	OFFICE	
MAYA	YARDS	SHED	OFFICE		SHED	WORKSHOP
BOB	OFFICE	WORKSHOP		SHED	YARDS	SHED

Making our farm shine

Red tags

While making things shine you will identify problems with equipment and items that are broken or not working correctly, such as:

- broken windows
- blown light bulbs
- damaged floors
- leaking pipes and cylinders
- broken inspection gauges
- gauges out of calibration
- broken or missing tools
- missing cleaning materials
- broken safety equipment
- broken wiring
- damaged switches
- blocked pumps/filters.

These are all maintenance items that should be fixed. It is therefore a good idea to allocate one person to record anything that you find needs some sort of action or repair. All problems should be recorded on a red tag log with a corresponding number. You can also attach a physical red tag (see figure 5.8) to the problem item so that the problem is easily visible and not forgotten. The tag number should correspond to the number on the record sheet. Once the maintenance has been taken care of, the item should be checked off, the date noted and the physical tag removed.

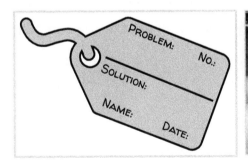

Figure 5.8: a red tag (left) and a red tag on gauge wiring (right)

Red tags can be any colour you like as long as they are bright and visible. They are a great way to visually identify problems, making it easy for team members to quickly locate the problem items.

The tag ideally should include a brief description of the problem, a log number that corresponds to the red tag log sheet so that a record can be kept, a brief description of a possible countermeasure (an engineering term for 'solution' or 'fix'), the name of the person who identified the problem and the date it was found. Don't make it too complicated as it will then put people off raising tags, as it will be 'too much effort' to fill one out.

Lean Farm team activity

Making your farm Shine

1. Go to the focus work area.

2. Clean everything in the work area including hosing, equipment, machines, walls, cupboards and benches. (Clean from the top down.)

3. Clean all the 'needed' items in the 'Sort' area.

4. Red tag all the problems you find and record them on a red tag log.

5. Move all the 'needed' items identified from the 'Sort' activity back to the work area ready to be 'Set in Order'.

You have now made your farm Shine — well done! Now that the work area is clean and tidy we can go back to the second 5S and discover how to put all the items needed in the work area into an order where everything has its home.

The 2nd S: Set in Order

Set in Order is about organising 'needed' items in the work area as efficiently and effectively as possible. Each item should have a very clear 'home' or set location so that everyone can find it immediately and knows where it belongs. Furthermore, the items and their set locations should be very visual.

A PLACE FOR EVERYTHING...AND EVERYTHING IN ITS PLACE

What and why?

Setting in Order is critical to ensure you have a well-organised, efficient workplace. It requires significant thought and planning. Most of us just put materials, items and equipment wherever we find a spot and don't put much thought into their ideal location. However, 5S means taking a step back. Ask yourself, 'Where is the best location for an item so that it is easily accessible to everyone who needs it?' and, 'How should an area be arranged so each item has a clearly visible home?'

Set in Order aims to simplify the process of using items and completing work tasks. It focuses on ensuring all items in a work area are arranged so they are close at hand to where they are needed, easy to use and clearly identified. Items should be labelled to make their designated storage location easily understood by everyone in the team and so that everyone can return items back to their home after use. High-use items should be located closer to the point of use, while less frequently used items should be further away.

Set in Order can make a big impact on our workplace, making processes simple and easier to complete, eliminating waste and frustration and improving productivity, safety and cleanliness.

How?

Once you have removed all the items that are not needed in the focus work area, you need to find suitable locations for all the necessary items. Arrange all the items so they are easy to find, use and put away. Establish a dedicated location for each item so that it is *always* in the same spot and could be found even if you're blindfolded.

Here are some steps to help you set things in order:

1. Determine how often each item is used—often, occasionally or rarely.

2. Arrange items, tools and materials in the order in which they are used.

3. Establish a dedicated spot for each item so it is easy to find, access, use and put away.

4. Think about how to optimise performance, ergonomics, comfort, safety and cleanliness when arranging items.

5. Place all regularly used items—such as procedures, instructions and specific tools such as spray cans and tail clippers—at their point of use (this reduces the time spent looking for them or retrieving them, meaning less waste).

6. Identify items and their 'home' or designated spot with labels or marked-out shadows or locations.

Examples of ways to arrange items are shown in figure 5.9.

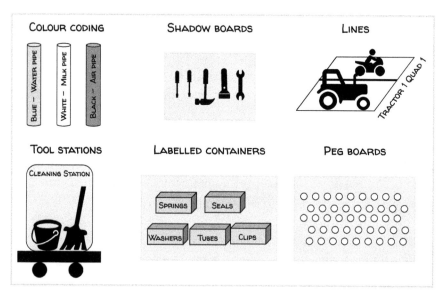

Figure 5.9: how to arrange items clearly and visually

Setting in Order the pump room with storage bins and dedicated locations

Key points to think about when carrying out this step are:

- Position items where they are needed—that is, at the point of use.
- Create a 'visual' work area.
- Communicate why things have been moved or located in a particular area.
- Make locations easy to find and see.
- Spend time planning the work area and item arrangement to optimise motion, efficiency, safety, ergonomics and cost.

Lean Farm team activity

The 5S number game: Round 3

We have now set our numbers in order. For round 3—the final round of the 5S number game—give each team member a printout of the following.

1	2	3	4	5	6	7	8	9	10
11	12	13	14	15	16	17	18	19	20
21	22	23	24	25	26	27	28	29	30
31	32	33	34	35	36	37	38	39	40
41	42	43	44	45	46	47	48	49	50

1. The numbers 1 to 50 are arranged in a grid.
2. Ask each team member to find and cross out numbers 1 to 50 in sequential order (i.e. 1, 2, 3, 4...).
3. Start the timer.
4. Record, on a flipchart, the time taken for the first person to have all their numbers crossed out.
5. Record the time when most other people finish counting.
6. Plot the first time on the graph on the flipchart paper and call it 'Round 3'.

Now discuss these questions with your team:

1. From the graph, can you see that on average it took significantly less time to sort the numbers compared to round 2 and round 1?
2. Why was it so much quicker?

3. Did setting in order help?

4. Can you relate this to areas of your farm: if everything was labelled and located with a designated visual home, how much easier would it be to find a tool?

Again this simple number activity demonstrates to your team the tremendous impact that setting the numbers in order has on the time it takes to do the activity. This is once again an exact reflection of our real world. If our farms were this clearly organised we would save considerable time every day. Now it is time to apply Setting in Order to your farm.

Lean Farm team activity

Setting your farm in Order

1. Go to the focus work area.

2. Identify where the best location is for all the 'necessary' items in the area based on ease/frequency of use.

3. Design the best way to store the items—for example, in a bin, on a bench, on a wall or on a shadow board (see figures 5.10, 5.11 and 5.12, overleaf).

4. Implement the storage solution: use pin boards, labels, colours and marking tape to make each item's home immediately apparent.

5. Place each item in its 'home' location.

6. Ensure each location is clearly labelled.

7. Take photos.

The key points of Setting in Order

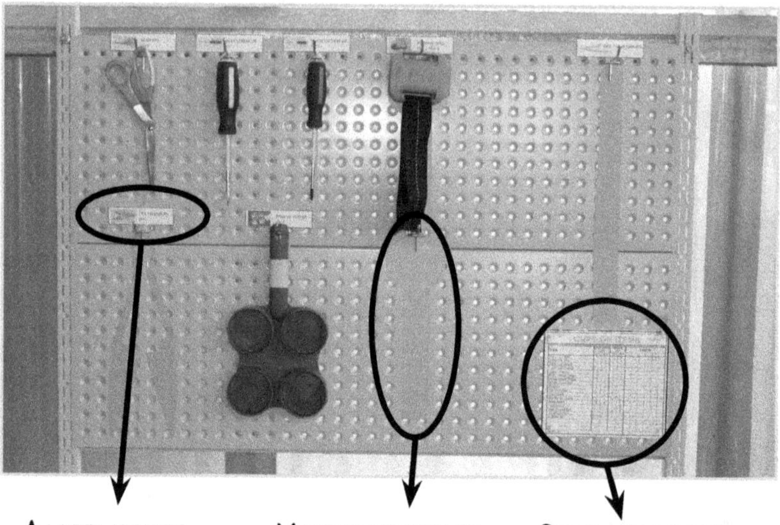

A LABEL NEXT TO EACH ITEM DESCRIBING WHAT THE ITEM IS, IDEALLY WITH A PHOTO OF IT

MARKED OUTLINE OF THE ITEM TO VISUALLY IDENTIFY WHAT TOOL SHOULD BE THERE AND WHAT IS MISSING

STANDARD LIST OF EACH ITEM IN AREA WITH A MINIMUM AND MAXIMUM QUANTITY SPECIFIED. IDEALLY ADD A PHOTO OF AREA TO REMEMBER WHAT IT SHOULD LOOK LIKE

Figure 5.10: the key points of Setting in Order

How Set in Order looks on our farm

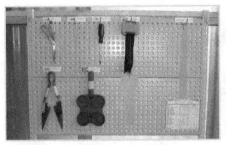

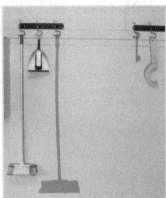

CLOCKWISE FROM TOP LEFT: CUPS–ON
SHADOW BOARD, SHADOW WALLS IN PUMP
ROOM, CUPS–OFF SHADOW BOARD AND
DRUG CABINET, STORAGE BINS FOR SHED
SUPPLIES IN PUMP ROOM, CLEAR LOCATIONS
AND LABELS FOR SHED SUPPLIES

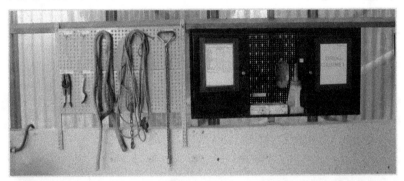

Figure 5.11: this is how Set in Order looks on our farm

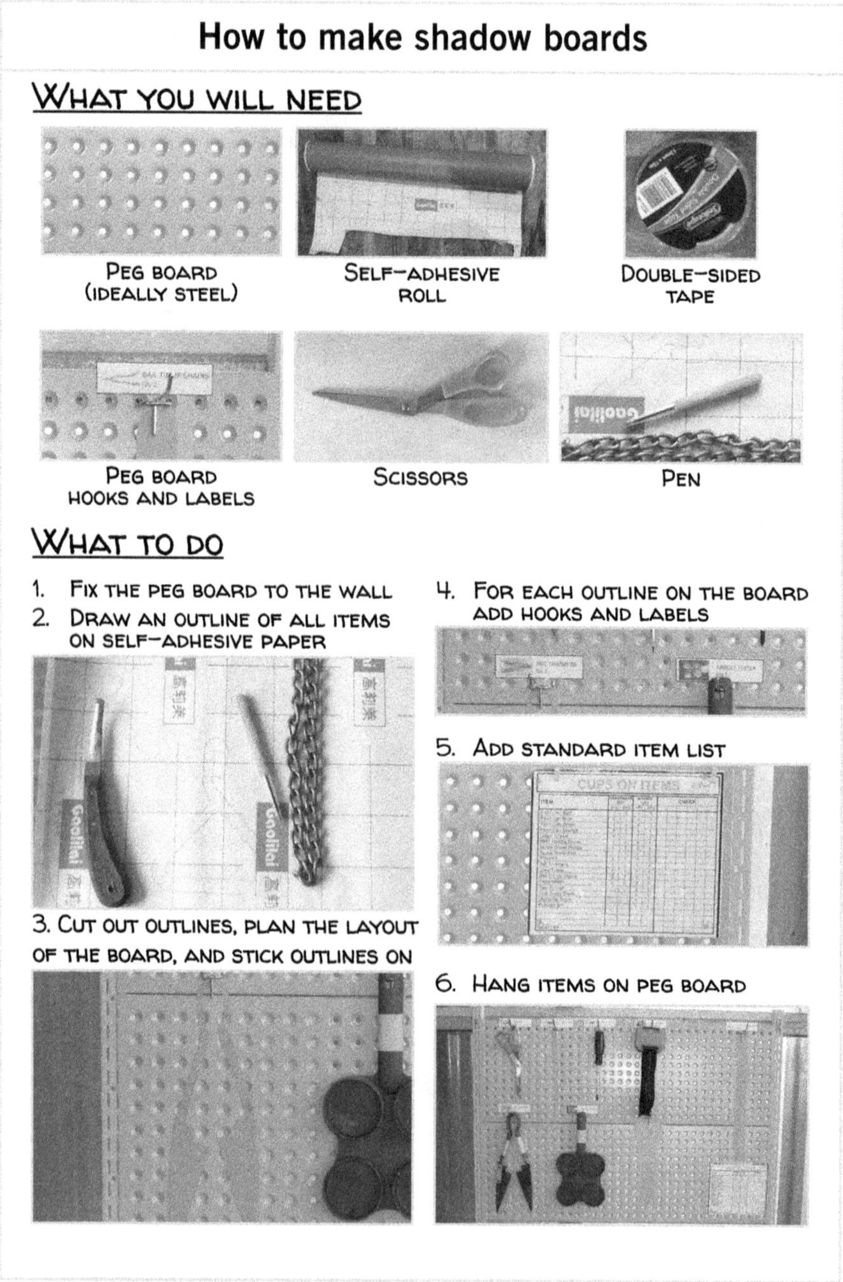

How to make shadow boards

WHAT YOU WILL NEED

PEG BOARD
(IDEALLY STEEL)

SELF−ADHESIVE
ROLL

DOUBLE−SIDED
TAPE

PEG BOARD
HOOKS AND LABELS

SCISSORS

PEN

WHAT TO DO

1. FIX THE PEG BOARD TO THE WALL
2. DRAW AN OUTLINE OF ALL ITEMS ON SELF−ADHESIVE PAPER

3. CUT OUT OUTLINES, PLAN THE LAYOUT OF THE BOARD, AND STICK OUTLINES ON

4. FOR EACH OUTLINE ON THE BOARD ADD HOOKS AND LABELS

5. ADD STANDARD ITEM LIST

6. HANG ITEMS ON PEG BOARD

Figure 5.12: how to make shadow boards like ours

By now you should be feeling very pleased with yourself (if you've completed the first three 5S steps) and even more pleased about how wonderful your farm is starting to look and how much easier it is to work in an environment that is sorted, clean and ordered. However, we can't stop there. If you want to ensure that your farm continues to be a tidy, well-organised, shiny farm you must put just as much energy and focus (if not more) into the last two steps of the 5S cycle. Let's look at step 4: Standardise.

The 4th S: Standardise

Standardise is a critical step in the 5S cycle. If you don't Standardise you won't be able to Sustain (step 5 of the 5S) any of the first steps you have completed. All the time you've invested in Sort, Set in Order and Shine will be in vain if there is no attempt to establish a clear standard of organisation and cleanliness in each area of your farm.

What and why?

Standardisation is about agreeing on the expected condition of an area and documenting that condition and the process required to maintain it. This standard then sets the guidelines for how each team member is expected to keep the work area. It ensures that the condition the particular work area is left in by each team member is not subjective but rather very clearly defined. By having a clear standard process for 5S, any discussion or debates about whether a work area is tidy or not is eliminated. The expected condition is black and white and it is very easy to refer to the standard work document and compare the 'actual' state to the 'desired' state to identify and address any discrepancies.

How?

There are several ways of creating a standard for 5S on your farm.

Daily 5S checklists

Checklists are an essential tool for standardising 5S. They should be placed in visible locations in the work area that they refer to. Checklists should identify the things that need to be done daily by all team members to ensure that 5S is maintained. This could be sweep the floor, wash machines, check shadow boards and so on.

The team member uses the checklist to tick off each item as they complete it. If any problems are identified during the 5S activity, the team member should raise them. The tasks on the 5S checklist should be done at an agreed standard time and frequency, which could be daily or weekly.

Visual standard documents

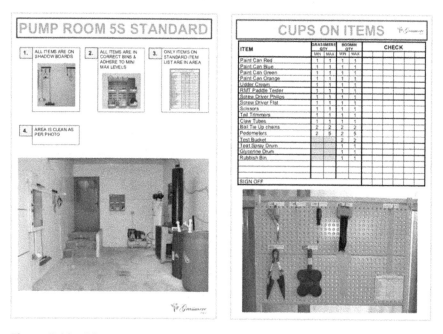

Figure 5.13: 5S standardisation documents on our farm

A visual one-page document showing the expected condition of each work area should be prepared and placed in that work area. The document should include photos of the work area and any key points to follow—like the one shown in figure 5.13. It can also include a list of the standard items that should be in that work area, where those items should be and the quantity of each. This way, anyone can go to an area, compare the area to the photo on the standard document and know whether the area is in an acceptable condition.

Lean Farm team activity

Standardising your farm

1. Agree on the acceptable standard of 5S for one farm area.

2. Take a photo of the desired condition of the area.

3. Develop a visual 5S standard document for the area describing what the area should look like.

4. Design a 5S checklist that the team can use to tick off specific 5S actions/checks to maintain the correct 5S condition.

The pump room before (left) and after 5S (right)

A Lean Farm example

No tools, no work

One farm where I ran my LeanFarm training program had a huge issue locating tools. They had several teams working across different farms all sharing the same set of tools. This meant if someone had to do a job, they often couldn't find the right tools and wouldn't know who had them. They would either waste time looking for the tools or have to phone all the other team members asking whether they had the required tool. This led to significant frustration, waste, lost time and cost in constantly purchasing new tools. After completing the 5S training, the farm immediately saw the benefit 5S could have on their team. They set up a tool shadow board in their workshop with all the tools on it. They also made ear tags for each team member with their names on them. Each time a person takes a tool from the board, they hang their ear tag on the board in place of the tool so that everyone else immediately knows who has taken that particular tool. Each team member has multiple ear tags in case they need several tools at the same time.

Standardising the 5S condition of the various work areas on your farm is essential. Without it your team will quickly fall back into their previous habits and your farm will look like it had never had any 5S done. This would be a real shame and will also possibly de-motivate your team. But standardisation alone isn't going to guarantee that you sustain your wonderful organised farm. This standard needs to be constantly reinforced and monitored to ensure that everyone is adhering to it and that the expected condition of the farm is maintained at a high standard. The final step in our 5S tool will help you to make 5S stick so that your farm will truly benefit from 5S in the long term and ensure that 5S becomes part of its operating philosophy.

The 5th S: Sustain

The final step in the 5S cycle is to Sustain. This means ensuring that all the previous steps in 5S are continually carried out and maintained in the long term so that improvements to the work, team, business and customer can be realised. This is the hardest part of the 5S cycle, but the most important. Without this, the first 4Ss are a waste of time.

What and why?

To Sustain something, you really need to change the mindsets of the team so that 5S becomes a way of life, a farm culture. It must become 'the way we do things around here' and needs to be embedded into the farm culture to ensure improvements generated from 5S can be maintained.

Sustaining is about:

- making 5S a part of your daily operation
- maintaining self-discipline and motivation to do 5S
- continuous improvement
- creating a culture that supports 5S and ongoing improvement and practice of 5S.

Sustaining your farm's 5S will provide long-term benefits to your business. It will make your farm a better, safer, more organised, productive and efficient workplace where employees will take pride in their work environment.

How?

There are many ways you can go about sustaining 5S in the workplace. Importantly, it needs to become the 'norm' on your farm: a part of the everyday process. This will require strong leadership and expectation-setting to ensure that the culture of the entire team shifts. *Everyone* in the team must be on board to sustain 5S.

Some things you can do to help Sustain your 5S efforts:

- Set a clear time each day to perform 5S daily tasks and ensure people do have this time available to do 5S.

- Make 5S a part of the standard process on your farm.
- Discuss 5S regularly at team meetings and at any opportunity to keep the focus on and reinforce its importance.
- Keep communicating the benefits, rules and expectations to everyone.
- Give rewards and recognition.
- Ensure satisfaction and enthusiasm.
- Address any 5S problems or concerns immediately.
- Address any deviations from the standard 5S condition immediately, before it slips too much.
- Use the 5S checklists and standard documents developed in step 4.
- Prepare 5S storyboards to show and remind people of the 'before and after 5S' condition of an area and the benefits to the team.
- Carry out regular 5S farm tours with the whole team to monitor all areas of the farm.
- Review 5S standardisation and promote 5S through seeing.
- Use the 'after 5S' condition as one of your business metrics and monitor it regularly at every team meeting.
- Measure the tangible benefits achieved by using 5S—time savings, savings in lost items, savings in scrapping expired products and so on—and give the team back some of these savings.
- Create 5S posters and hang them around the farm to keep 5S front of mind.
- Make 5S one of the criteria or expectations that form part of the performance development process and discuss each individual's compliance.
- Ensure all team members are self-disciplined and hold each other accountable.
- Ensure 5S is *everyone's* responsibility.
- Lead to continuously role model the right 5S behaviours—never walk past an unacceptable condition and ignore it.

Your leadership will be critical to the successful implementation of 5S and all the other Lean tools on the farm. There are four key actions that anyone wanting to introduce a change successfully needs to diligently practise as a minimum. These four actions will help your farm to sustain any Lean tools introduced, including 5S.

Lead by example

Show your team that you are serious about 5S. Never walk past rubbish without picking it up. Walk the talk.

Don't ignore

If something isn't right, address it immediately. Never ignore a situation or turn a blind eye. This will just make everyone think you aren't taking it seriously.

Communicate

Constantly communicate, communicate, communicate. Keep talking about 5S every day. The more you talk about it, the more people will be constantly thinking about it. It will drive the message home.

Audit

Prepare audits of the 5S condition that are run by farm managers as well as your team on a regular basis. This way everyone is accountable and disciplined. See figure 5.14, overleaf, for an example.

* * *

If you would like some visuals around your farm to give your team a constant reminder about 5S and what it means, you can use the poster reproduced in figure 5.15 (overleaf). It will help to keep 5S front of mind and hopefully start to drive the right behaviours.

DAIRY SHED 5S AUDIT SHEET

MONTH: JULY   **WHO CHECKED:** MAT   **DATE:** 30/07/18

AUDIT CHECK ITEM	OK	NOT OK	COMMENTS
PUMP ROOM			
Area is clean and organised as per standard work	✓		
Only items listed on standard item list are present in the area		✓	Thistle grubber and other tools are in pump room
All items are in the correct place	✓		
No items are missing		✓	The yard broom is not there
CUPS ON AREA			
Area is clean and organised as per standard work	✓		
Only items listed on standard item list are present in the area	✓		
All items are in the correct place		✓	Tail scissors in wrong spot
No items are missing from the shadow boards	✓		
CUPS OFF AREA			
Area is clean and organised as per standard work		✓	Floor area has empty buckets and left over metabolics bags
Only items listed on standard item list are present in the area		✓	Pipe wrench has been added to shadow board
All items are in the correct place	✓		
No items are missing from the shadow boards	✓		
SHED GENERAL			
Dairy shed floors and walls are clean		✓	Walls need better hose down
All hoses put away correctly	✓		
Platform is clean		✓	Displays need wipe down

Figure 5.14: a worksheet for a 5S/visual management audit

5S on our farm

1. **S**ort

2. **S**et

3. **S**hine

4. **S**tandardise

5. **S**ustain

Figure 5.15: a 5S poster

Lean Farm team activity

Sustain your farm

1. With your team, brainstorm and record the following on a flipchart:

   — What will you do to Sustain 5S?

   — What are the barriers/challenges for your team in sustaining 5S?

   — How will you overcome these challenges?

   — How will you hold each other accountable?

2. Design an audit sheet that the team can use to audit the 'after 5S' condition of one of your farm areas.

3. Decide how often audits are required and how responsibilities are to be shared among your team.

Designing and planning for 5S

One of my greatest bugbears in farming is that a lot of our farm facilities and equipment are designed, built or installed with not a lot of consideration given to how easily they can be maintained or kept clean. It often appears to me that some of those involved in designing dairy equipment have never worked on a dairy farm. There are inconvenient, hard to get to corners that are difficult to hose out and just end up collecting rubbish; concrete that is prone to water collection and becomes super slippery; pumps that have no proper covers on them so they collect dirt; awkwardly positioned rails, fences and yards that make them hard to clean; drainage located away from taps, allowing water to flood and run across floors before finally making it to the drain. I could reel off multiple examples of poor design from our farm that are just impractical, make life hard and are not conducive to keeping a farm in good 5S condition.

However, there is hope. Now that you are aware of 5S and how important it is for your farm, you can think about it every time you plan on installing new facilities, designing a new work area or doing a new layout. This way you can design your work areas so that they aid 5S or are inherently, by design, in 5S condition from the start. So remember to design for:

- *Sort*—find ways to prevent items from accumulating in the work area by removing hidden corners from designs/layouts or unnecessary cupboards/rooms/spaces
- *Set in Order*—make it difficult or impossible to put things back in the wrong place or to add extra things. Limit shelving and bench space; use only peg boards; set up dispensers suitable for only specified items
- *Shine*—prevent work areas from getting dirty in the first place by designing facilities and work areas so that there isn't a possibility of contamination. Correctly positioned drains; good maintenance systems; proper fittings; suitable and effectively positioned cleaning facilities such as hoses or water cannons; and easy-to-access areas for cleaning instead of awkward corners and shapes will stop dirt, excrement and other unwanted substances from appearing or accumulating.

Summary

5S is one of the key foundations of Lean thinking. It is a very simple, quick and cheap way of making some immediate improvements on your farm. 5S alone can have a tremendous impact on your farm's efficiency and the way you work. It will save time and money and prevent frustration among your team. The difficult part is to sustain it.

5S is not just a one-off activity. It must become a part of your business culture: the way things are done and what is expected of everyone. Everyone is responsible for doing and maintaining 5S.

5S can help you to:

- improve key farm metrics such as animal health, quality, safety, production and cost
- create a well-organised, well-run farm
- demonstrate that your farm is under control
- remove unnecessary items (by re-deploying them elsewhere if needed)
- create a clear standard of what is expected
- continue to improve your farm
- improve morale and working conditions
- save money on expired or lost items/materials
- create a safe and professional farm environment.

Importantly, your team must be involved in the 5S activities you do. If they are involved they are more likely to sustain the condition as they have done the hard work and know what the workplace looked like before 5S.

Lean Farm team quiz:
5S

1. What is a red tag used for?

2. What does 5S stand for?

3. Which step in 5S is the hardest to do?

4. What are five benefits of 5S?

5. What type of documents should you create to help make 5S a part of your everyday work?

6. What are the four piles you should organise things into when doing the first step of 5S?

7. Who should be responsible for 5S?

8. What are the four leadership actions that will help sustain 5S?

5S for calving—a standard calving box with an item list

Lean Farm action plan:
5S

This action plan is included to provide you with some simple guidance. It is aimed at giving you a little bit of motivation and direction if you need it. Of course, the idea is for you to do as much as you can to gain the maximum benefit out of 5S on your farm.

1. Introduce 5S to your team.

2. Identify at least two areas on your farm to carry out 5S.

3. Take *before* photos.

4. Complete the five 5S steps in each work area.

5. Take *after* photos.

6. Develop the necessary documentation to Standardise and Sustain 5S.

7. Set a time for 5S every day or week.

8. Include 5S as part of your performance review process.

9. Include 5S discussion in every team meeting for ongoing reinforcement and focus.

I think your team will get a lot out of 5S and will enjoy implementing it. It is the tool that gives the biggest immediate visual impact on the farm and gives people an immediate sense of achievement and satisfaction. The next tool we will talk about is visual management. 5S is closely linked to visual management as a lot of 5S is about making your workplace and the things in your work areas more visual so they can be easily found.

Chapter 6

Visual management

Visual management (VM) is another fundamental foundation of a Lean business. Like 5S and the 8 wastes, it is a very simple concept that is low cost, easy to implement and can bring considerable benefit to your business. In fact, one of the tools we will discuss in this chapter is probably the most valuable tool for farms to focus on and will help you achieve the right results for your business. While much of what we talk about in this chapter is again common sense, like the other tools its success relies heavily on changing the mindsets and culture of your team (and we look at ways to do this in part III). If you do this successfully, VM can truly turn your business around or help you achieve your goals effectively.

What is VM?

Visual management underpins the way a Lean business manages its people and its operations. It is a very powerful Lean technique that is proven to enable any business to achieve their goals and targets effectively. If you want your farm to achieve its targets, let alone exceed these targets, I would highly recommend you start using visual management. I can almost guarantee that if you apply this tool correctly in your business the results will be incredible.

Visual management is an approach that allows everyone in a team or business to quickly understand what's going on. It is about the visualisation of key information by and to all so that everyone has access to the same vital information, is involved in the information and knows what the information means. Then all management and decision making on the farm is based on and around the visualised information. The visual information is used as the focal point for the team. Visual management is a critical support tool for all the other Lean tools and principles including 5S, standard work, total productive maintenance and process mapping. In the words of the man considered to be the father of the Toyota Production System, Taiichi Ohno, '**Make your workplace into a showcase that can easily be understood by anyone at a glance. Problems can be discovered immediately and everyone can initiate improvement plans.**'

The majority of people are visual communicators. This means that if someone receives instructions or information visually, it is much easier for them to understand what is needed. If you try to explain something to someone in words and sentences, more often than not the message is lost in translation.

Visualisation of information is particularly important if your farm has a multinational team. A simple picture, visual instruction or photo can make something much clearer so that a job is done right the first time (rather than having to read a page of writing or listen to someone's explanation).

Furthermore, when people have access to information they are able to *understand* better, resulting in more informed, accurate decisions and actions. This leads to better results.

Good visual management enables everyone to immediately see deviations from the optimum state of work and working, and to enable immediate corrective action. It should not need any interpretation—interpretation takes time and can lead to mistakes. The time it takes to react to a visual signal or information is important—successful visual management in the workplace will ensure accurate understanding of what is required, an immediate response and that the right action is taken by everyone. Good visual management will also improve the speed and accuracy of our response to problems (see figure 6.1).

INFORMATION = ACTION = RESULTS

Figure 6.1: impact of visual management

Lean Farm team activity

Using visualisation on your farm

Think about your farm. Identify at least five things that you think could be better visualised on your farm. Write these down.

Discuss your ideas with your team.

Which paddock? Our farm paddocks have no paddock numbers. Have we had paddock mix-ups? Of course...and I don't blame the team.

Is it in your head?

Traditionally, farmers are terrible at sharing, let alone visualising, information with others. Most farmers have decades of strategies, information, planning, knowledge, processes, numbers and results stored in their heads. Often, very little of it is made available in an easy format to others who need it. Unfortunately, most of us are not telepathic and can't read other people's minds. It is very difficult for others to do what you want, when you want it and how you want it if they don't understand exactly what is expected and needed and the reasons or background to something.

What is the wider or longer term plan/schedule? What are the priorities? When are key farm activities such as herd testing or mating happening and how? Why does something need to be done? Who is doing it? When should it be done by? What is the process and expectation? What are the targets for the farm/team? How are we going against these targets? All of this is important information that everyone in a team needs to know in order to work efficiently and effectively.

It is no wonder that things go wrong, nothing gets done, results aren't achieved and people get cranky when you have no information and rely on what's in someone's head and what they remember to share with you.

Traditional farmers tend to inform their team of things as they remember them or at the time that something needs to be done (sometimes last minute). The information comes out of the farmer's head and is delivered to the team on an ad hoc basis during the work day. There is little structure or visualisation of key communication and information. Lean farmers are different. They use visualised information and communication in a standard approach to ensure that nothing is hidden in people's minds. It is all made available, visual and transparent for the whole team so that everyone is aware of what is going on and involved in the decisions, problem solving and discussions where needed. Everyone therefore knows what's coming up, what the plan is and why things need to be done, when and by whom. There aren't any surprises. Everyone can work together, towards a common purpose with clarity and understanding. See the difference between a traditional and a Lean farmer in figure 6.2.

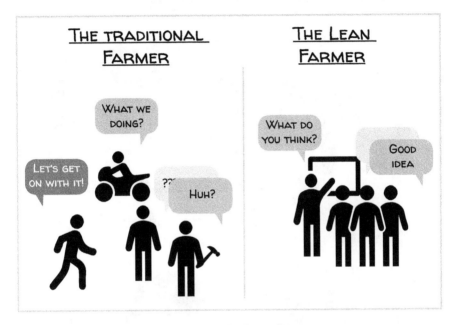

Figure 6.2: the traditional farmer vs the Lean farmer

The two key elements of visual management that I will discuss in this book are *visualisation* and *visual management boards*.

Visualisation

Visualisation, or visual communication, is all around us. It is an inherent part of our lives and helps us to communicate or share information, ideas or messages visually.

It helps us to understand a situation or message quickly and to make immediate decisions without having to go and ask someone a question or get out a 100-page manual to work something out. It is a critical aspect of our everyday life and without it our daily life would be pretty difficult. Visualisation through visual imagery has been used for thousands of years and it still remains a powerful tool today in all industries. There are many forms of visualisation, with the most obvious being visual signals such as those that follow.

A PICTURE IS WORTH A THOUSAND WORDS.

IMAGINE IF THERE WAS NO VISUALISATION...

... AT THE AIRPORT...

... OR ON THE ROADS!

HOW COMPLICATED AND UNCLEAR WOULD LIFE BE?

CAN YOU IMAGINE THE TYPE OF PROBLEMS YOU WOULD HAVE?

IF YOU CAN'T IMAGINE LIFE WITHOUT VISUAL COMMUNICATION, HOW CAN YOUR BUSINESS OPERATE WITHOUT IT?

Visualisation should be so clear that it doesn't need any interpretation — it should make you do the right thing just by looking at it. When you stop at an intersection, how do you know when you can accelerate again? Do you have to wind down your window and ask the driver next to you? No. The traffic light turns green and you immediately know that you can accelerate. That is an example of simple, effective visualisation.

How can visualisation be used?

Visualisation can be used in many areas to help us do our work more effectively and efficiently.

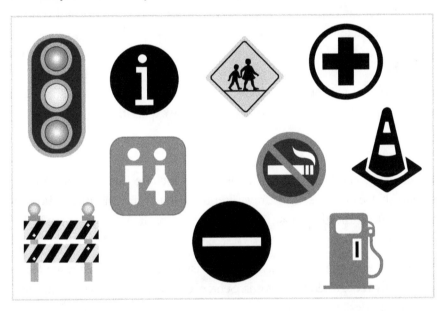

Visualisation is all around us and we rely on it every day.

Quality

Gauges marked in red and green to show if the dial is within specification or outside of it can help to immediately see if equipment is functioning correctly. If we had no visualisation, either you would need to remember the correct specifications or you would need to find them in a manual. You would also potentially have a problem, such as the incorrect temperature of your hot water, for some time before anyone noticed it. Visualisation like this can help the team see a problem immediately without needing to be an expert on the particular piece of equipment.

Production

Production can be monitored using visualisation so that if there is a problem it is immediately displayed and action can be taken. It helps to track whether you are on target.

Maintenance

Maintenance can be simplified by using visualisation. Colours to represent areas where machines or equipment need to be maintained can help the team to navigate complex machinery. Labels on dials, switches and control panels can make it easier to use.

Safety

Safety is a vital part of our everyday work life and visualisation is important to help establish, inform and reinforce any safety expectations and standards.

***

Examples of good visual systems or signals include illumination (a light signal) that can be used to draw attention to a situation either before it arises or to indicate that an abnormality has occurred (for example the petrol light coming on).

Symbols are another good way to create visual instructions. They are more universally recognised than words. Sentences take much longer to interpret than a well-designed symbol.

Standard colour coding can also be used as a powerful visual signal to reduce the time it takes to interpret a situation and therefore support a fast reaction. Red and black leads to jump-start a car are a good example as are petrol stations, which use different colours to represent the different fuels.

A Lean Farm example

When poor use of visualisation costs big time

I was told a story about a farm that had put cows into the wrong paddock for grazing. Unfortunately, it wasn't just about cows grazing the wrong paddock and loss of grass. The paddock that they had been put into had just been sprayed with weed killer. The cows then went on to be milked and the milk went into the Vat. By the time someone had realised that the cows had been grazing a sprayed paddock, the milk had already been picked up by a milk tanker and therefore affected a whole tanker full of milk (and wasted several other farmers' production).

This is a true example of where visual management could have prevented an unfortunate event. If the paddocks had had huge, bright, yellow signs reading 'paddock 50'—when the cows were supposed to be in paddock 35—the person setting up the paddocks would have looked twice and not made the mistake of putting them in paddock 50.

Furthermore, if there had been some visual indicator such as a big red tape/sign blocking the paddock after it had been sprayed, again the person setting up paddocks would have immediately stopped and understood that the paddock was out of bounds and had just been sprayed. Two simple visuals that would only cost you $50 would have saved a wasted vat of milk (not to mention the whole tanker). Not only that, but also the time and money spent harvesting the milk—grass, labour, materials, feed, electricity and so on.

Visualisation on the farm

Our farms are full of examples of visualisation (see figures 6.3, 6.4 and 6.5 overleaf). In fact, we are masters at creating visual signs or signals to help us manage our farms more effectively. Next time you spray paint a cow, think about why you are doing it. Does the colour stand for something? Yes, a cow with red on its udder usually means 'mastitis' or problem cow. The cow is painted so that everyone working on the farm can immediately identify it as a problem cow and monitor it. Ear tags, equipment control panels, tank level indicators, meters, safety signs, farm maps, animal health management posters and grazing plans are all common examples of visualisation that exist on our farms. These help our teams to identify problems, or obtain the right information they need and take the right action quickly.

Figure 6.3: safety signs

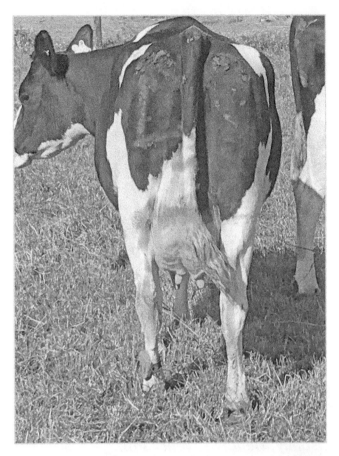

Figure 6.4: red paint on mastitis cows and cow ear tags

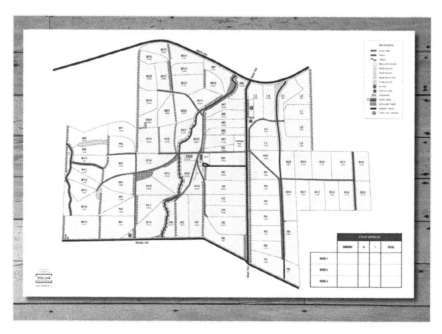

Figure 6.5: visualisation of farm paddocks

Source: Courtesy of VizLink

Lean Farm team activity

How visually friendly is your farm?

Walk around your farm and look out for any visualisation. Use the visual management audit sheet as a guide (see figure 6.6).

1. What are examples of good visual management that you can see on your farm? (Take photos.)

2. What are examples of opportunities to apply visual management on your farm? (Take photos.)

3. What five actions can you take to improve visualisation on your farm?

4. Discuss what you have found with the team.

FARM VISUAL MANAGEMENT AUDIT SHEET

MONTH: August   **WHO CHECKED:** Steve **DATE:** 19/08/18

AUDIT CHECK ITEM	OK	NOT OK	COMMENTS
HEALTH & SAFETY			
SAFETY SIGN DISPLAYED	✓		
EMERGENCY EXITS, KIT AND PLAN DISPLAYED	✓		
VISUAL TRACKING OF SAFETY IN PLACE		✓	SAFETY CROSS NOT UPDATED
TEAM VISUAL MANAGEMENT BOARDS			
VISUAL BOARDS PRESENT AND ACTIVELY USED & UPDATED	✓		
INFORMATION IS RELEVANT, ACCURATE AND LIVE		✓	CALVING METRICS NEED TO BE UPDATED
KPIs/ ACTION PLANS ARE VISUALLY RECORDED, MONITORED	✓		
PROCESSES			
STANDARD WORK INSTRUCTIONS, PROCESSES EXIST AND ARE VISUAL		✓	NEED TO ADD LATEST PROCESS ON CALVING INTO FOLDER AND ON WALL
VISUAL DIAGRAMS USED TO EXPLAIN PROCESSES/EQUIPMENT	✓		
FARM VISUALISATION			
PADDOCKS ARE LABELLED WITH CLEAR SIGNS OF NUMBER ON EACH PADDOCK		✓	SOME PADDOCKS DON'T HAVE LABELS YET
SILOS ARE LABELLED VISUALLY	✓		
CALF PENS/FENCES/GATES ARE NUMBERED AND COLOUR CODED		✓	PENS 1, 5 AND 6 HAVEN'T GOT CLEAR SIGNS AND COLOUR MARKINGS FADING
FARM VEHICLES ARE LABELLED AND PARKING LOCATIONS IDENTIFIED AND LABELLED	✓		
SHED VISUALISATION			
CONTROL PANELS, PLATFORM DISPLAYS ALL LABELLED/COLOUR CODED	✓		
TANK MIN/MAX LEVELS AND TEMP DISPLAYED/VISUAL	✓		

Figure 6.6: a farm visual management audit sheet

Farm problems resulting from poor visualisation

While there are some great examples of visualisation on our farms, there are also plenty of opportunities to be better at it.

How many of you have had cows put in the wrong paddock? Or had a delivery truck show up and dump something in the wrong place? Or had calves mixed up? What about herds being mixed up or in-calf cows mixed with others? Some of these problems can be very costly, right? And they most certainly are a nuisance, wasting somebody's time and energy to fix them. Most of these problems could be prevented if we had better visualisation on our farms. Having your farm maps visualised, for example, can be a good starting point and make a big difference to your team. These maps can help you visualise important information about your farm such as where paddocks are, what the paddock numbers are, where the electric fences and water pipes run and the size of paddocks. This can help to avoid many basic problems or mistakes.

Here are just a few very common examples of on-farm problems often caused by poor visualisation.

Cows in the wrong paddock

Have you ever had cows put in the wrong paddock? Or what about a team member phoning you to ask which paddock was number 47 and you having to try to describe it over the phone?

Having a clear visual farm map and better visual numbering of paddocks that match your farm map can help anyone immediately see whether they are in the right paddock without having to make second guesses or call someone. Having copies of farm maps that your team or suppliers can take around with them can also help to ensure a person is in the right paddock.

Tape gates broken

It is a cold, rainy (maybe snowy), foggy winter morning. It's dark and you need to go and bring cows in. There are thin white tape gates across various races. You drive through one, breaking it and almost knocking yourself off your bike ... Has this happened on your farm? Are you surprised? Not only are you now going to have to waste time doing unplanned maintenance (repair) on your tape gates, but you could have had a serious accident.

Having highly visible flags on tape gates is a simple way to ensure that they are visible even in poor weather conditions.

Milk taken from/put in the wrong Vat

Whoops … this is unfortunately quite an expensive mistake. Perhaps a team member has put red milk into the Vat, or your milk supplier has come along and picked up colostrum milk destined for calves instead of the correct milk from the right Vat. This can easily happen if there is not good visualisation of Vats and some 'poka yoke' systems in place (we talk about these systems in chapter 10). A mistake like this can not only have big consequences for your farm, such as bad quality grades and cost, but it can also potentially affect the whole supply chain upstream if it isn't discovered in time.

Meal in the wrong silo

One of our suppliers coming to fill up our silos. Each silo contains something different. Can you spot a problem? These guys have seen hundreds of silos just like this. How are they going to remember which one to fill?

Have you had suppliers accidentally put the wrong feed into a silo? It has happened to us. And really, it's no surprise. The contractor has most likely just driven to several farms that day delivering feed—all of them with similar-looking silos. Unless there are very clear visual signs describing each silo, the contractor is hardly going to remember which one is which on your farm when they have perhaps hundreds of farms they supply. When a problem like this happens, we need to question how we can visualise our work environment better to prevent the same thing happening again.

Also, it's good to note that there is a big difference between having signs stuck up everywhere and true visual management that is actually *visual* and effective at prompting the right behaviour and outcome. Just because we have a heap of signs up, it doesn't necessarily mean they're visual and effective at communicating the right thing. Therefore, always ask, 'Is this really visual to everyone? Will it guarantee the right action?'

A Lean Farm example

Silo mix-up

I visited one farm where they mentioned that they still had the wrong feed put into the wrong silo even though they had clear signs on each silo indicating its silo number and what feed should go into it.

However, when you actually looked at the silos, the visualisation was not clear at all. There were numbered signs as well as signs saying 'meal', 'PKE' and so on. However, these did not correlate.

Furthermore, the signs were sitting at the bottom of the silos where they were hard to see. If a truck driver is reversing in to deliver feed into a silo, and they are sitting high up in their truck, how are they going to see a sign that is on the ground behind their truck? They won't—therefore it isn't visual enough.

If a mistake like this has happened despite some visuals being available, you need to look at your visualisation and question whether the process or work area is *truly* visual.

Jaya spray painting cows to visually identify them

Lean Farm team activity

Visualisation

With the whole team, brainstorm and discuss some recent problems or mistakes that you have had on your farm. Think about:

— what the problem was

— why it happened

— whether it was visual enough

— whether better visualisation could have
prevented it.

You're not a lone ranger

We have to remember that we don't work alone on our farms; we are interacting with the outside world on a daily basis. We have milk suppliers, maintenance guys, tradies, contractors, consultants, suppliers, vets and all sorts of other people coming to our farms daily. These people don't just visit our farm—they have been to probably 100 or more other farms before yours and they are most certainly not going to remember every detail of your farm. Therefore, creating simple visualisation on your farm can help everyone navigate your farm quickly and easily, preventing disruption to you and your team, inefficiencies, frustrations and, importantly, costly mistakes.

Here are some examples of where visualisation can be implemented on a farm:

- installing paddock numbers on paddock gates/fences
- setting colour markers in paddock to identify location for breaks for easier set-up
- putting labels/signs on silos
- labels on diesel and petrol tanks
- putting labels and red covers on Vats
- visualising grazing plans
- attaching labels to control panels describing what each button or switch is for
- using visual cow identification
- creating and displaying farm maps
- visualising where water pipes and taps are and which taps connect to which water pipes
- visualising your paddock water pipe layout
- identifying sick/mastitis cows visually with colour
- using colour/numbers for calf pens and associated fence panels/gates
- signs identifying farm safety/hazards
- displaying visual farm processes
- visual signs for safety requirements
- 5S/visual locations for consumable items and stock levels (treatments, gloves, chemicals etc.)
- animal health.

Visual management boards

The other important element of visual management is what is sometimes referred to as visual management centres or boards (VMB). These are essentially walls, boards or rooms displaying all the critical information related to your farm in one area. The information displayed helps a team to have transparency and understanding about what is going on so that everyone is on the same page at the same time. It helps to ensure that everyone in a team is aware of the key issues, priorities and actions required. Problems can be identified and discussed immediately in real time and accurate decisions can be made.

Having a good team visual management board can help your farm develop a high-performing team and achieve farm targets successfully. The visual management board will usually display information such as your weekly plan, any actions, your priorities, your grazing plan, any maintenance required, general team communication, your production and quality results and other important metrics, and anything else that is important for the team to be able to work effectively. These boards are usually in the form of whiteboards that are broken into various sections to display the required information. Importantly, the visual management board or wall will be the focal point of all communication, discussion and decision making for the team.

Note: Even if you are a one-person farm, don't ignore this section — visualising your information on a wall, even if it is just for yourself, will help you to stay focused, on top of your business and to achieve your goals much more effectively.

TRADITIONAL: 'WHAT GETS MEASURED GETS MANAGED'

LEAN VISUAL ENVIRONMENT: 'WHAT GETS MEASURED VISUALLY GETS MANAGED IN REAL TIME'

Visual management board rules

While most of us may think we already have some sort of visual management board on our farm, there is a big difference between a whiteboard with some stuff written all over it and a proper, effective team visual management board. To get real benefit from this Lean tool, your visual management board must fulfil these three rules:

- *Be clear and visual (crystal clear to anyone).* Your boards should be very easy to understand. Anyone from outside your farm should be able to look at your visual boards and understand what they are all about.

- *Grasp the situation in **3 seconds**.* What is the target condition (desired state), what is the actual condition (how are we really doing today) and how big is the gap between actual and target. You should be able to look at your board and understand the current situation in 3 seconds. If it is truly visual, that's how easy it should be. Using red, green and amber colour coding is a great way to immediately spot problems.

- *Be live and updated.* There's no point having team visual boards if no-one updates them and the information on them is a month old. To make visual management highly effective, the boards need to reflect the current state of the farm. The team needs to ensure that all information on the board is up-to-date, relevant and accurate. The boards need to be used regularly and form the centre of all team and management communication and decisions.

What do visual management boards look like?

Visual management boards can really be anything you need them to be as long as they display relevant, accurate, live information and are used by the team. The team or farm needs to design these to suit your farm needs and budget.

Visual management boards can be very basic to start with, simply visualising perhaps your grazing plan, together with some key actions, maintenance required and a weekly plan. This would be a good starting point.

Once your team has got used to the visual approach, you can begin to introduce other elements to your visual management boards such as problems, metrics, graphs, season plans, continuous improvement ideas and so on.

Most visual management boards consist of one or more whiteboards with certain headings and sections marked out on them. In more recent times, however, visual management boards are becoming more sophisticated, with touchscreen and electronic display technology being used instead.

On our farm we have two levels of visual management boards. We have a team visual management board set up at each of the dairy sheds. This is shown in figure 6.7. These boards were designed by the team and are owned by the team. They include all the information the team wants and needs to be able to run the farm. This includes a weekly plan, safety, any problems, maintenance items, supplies needed, action plan, season plan, grazing plan and all our key metrics.

Figure 6.7: our team visual management board for the Grassmere shed

We also have a next-level-up management team visual board, which is used by the managers to report back to Mat. This board contains a higher level of the information obtained and cascaded up from the team boards. It includes graphs of key metrics (showing target vs actual condition), any escalation items, key pasture management points (pasture cover, round length) and a weekly action plan for the management team.

The boards are used for the weekly team and management meetings. They are the focal point of all discussions, decisions and actions. As the boards are situated in the team meeting rooms and are visual, they are seen by the team every day. This keeps the information front of mind.

Lean Farm team activity

What to put on a visual management board

With the whole team, brainstorm information that would be helpful for the team to have on their visual management board. Think about:

— type of information (e.g. actions, grazing)

— how the information should be displayed/
visualised (e.g. picture, table, graph, free text).

Designing an effective team visual management board

Visual management boards will only help your farm and be successful if done correctly. The most important thing is that your team is involved in designing them. This will make it more likely that the team will own the boards and use them effectively. If this isn't done, the boards will just become a scribble on a wall. Some key points to consider when setting up your boards is to:

- involve the whole team
- let the team design and agree to the content and layout
- make it relevant (only have information that is useful to the team)
- include clear roles and responsibilities
- define the frequency, process and owner for updating the board
- have a rotating roster
- ensure the information is updated and live
- make it visual: use graphs, pictures, colour
- keep it simple

- make sure the board drives a focus on actions (including follow-up)
- ensure there is continuous improvement — don't try to make this your perfect board; just get something set up and evolve it over time as the team start to use it and become more familiar with what they want.

What to have on your board

Here are some ideas for information to include on your team visual management board:

- key farm metrics
- production results
- quality
- maintenance items
- materials/supplies needed
- grazing plans
- animal treatments required/ cows to watch
- problems/risks identified

- actions for the day/week/ month including clear responsibility and due date
- priorities for the team and farm
- skills within the team and any training needed
- holiday/leave plans
- special projects
- safety issues including near misses, risks, hazards or incidents.

A Lean Farm example

Designing our team visual management board

Step 1

The first step in designing your team management boards is to get the whole team together and, using butchers paper, a whiteboard or sticky notes, brainstorm all the things that the team would like to have on their board. This is an example of our brainstorm. We put a piece of butchers paper on the wall and wrote down all the things the team came up with.

(continued)

A Lean Farm example *(cont'd)*

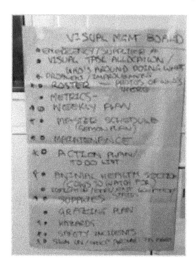

Step 2

The next step is to sketch up what the whiteboard would look like using the information from your brainstorm. We simply used bits of paper and butchers paper to draw it up and displayed it on a wall so we could visualise what the final board would look like. Once you see it on a wall, the team can make changes to the content, design and layout several times until they are happy.

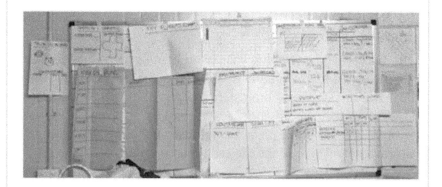

Step 3

We then designed the templates using Word/Excel so we could get a better picture of what the final boards would look like. This makes the board look real and you can very quickly identify any tweaks needed and also how the layout and size will work with the space available. Don't get too hung up over making this perfect right away. Your board won't be perfect—it will evolve over time.

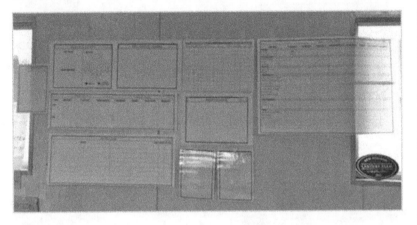

Step 4

The last step is to set up your templates on a whiteboard so you can start using your board. Once your final board is up, and you start to use it, you will identify things to improve and you can evolve your boards over time. This is continuous improvement.

(continued)

A Lean Farm example *(cont'd)*

Here are our team visual board before and after photos.

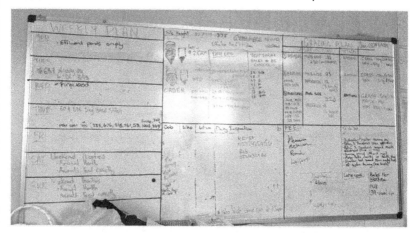

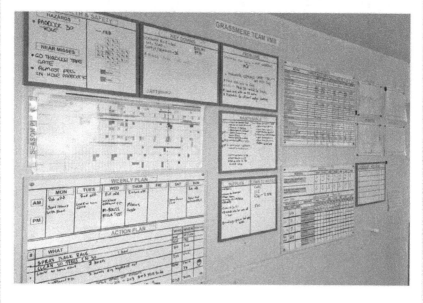

Setting up your visual management board

There are a number of options you can take in setting up your team's visual management boards depending on your budget. If you have money to spend, there are professional companies that can design customised whiteboard templates for you and manufacture your whiteboards with the template on it. This will create a very professional-looking, high-quality and long-lasting board. It also makes the process quick and easy (see figure 6.8).

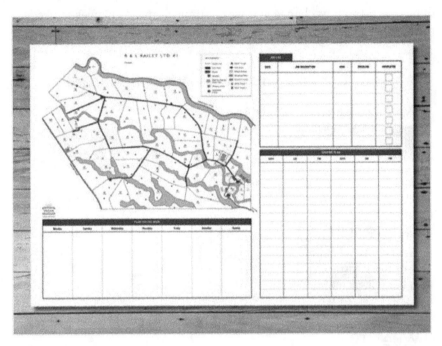

Figure 6.8: companies such as VizLink in New Zealand make professional customised visual whiteboards like this one. They can work with you to design the right board for your farm. This makes the process easy.

Source: Courtesy of VizLink

If you want something on a low budget, you can buy whiteboards in the size you want (you can get cheap ones online or in low-cost shops) and then use whiteboard grid tape to mark them up (best to buy this online also as it's much cheaper—the German stuff is very good). Gridding whiteboards is time consuming but it's a good option that is affordable and you will still get a professional-looking board that will last a long time. Ours has been up for a year and a half and is still as good as gold. I have even set up boards like these for executive teams at some big companies and they serve the purpose well.

Of course, if you want to just get started, get a whiteboard and you can simply mark it out in permanent marker pen.

Don't forget, your whiteboard won't be perfect first time. Ours most certainly wasn't. Over the time that we have had ours up, we have identified many things to add to it or change and we continue to make changes as needed. Figure 6.9 shows how to make a visual whiteboard.

Figure 6.9: making your own visual whiteboard

Benefits of team visual management boards

As mentioned earlier, this is probably one of the biggest improvements you can make to your farm to ensure better communication, engagement, ownership and results.

The list of benefits is significant, but here are just a few:

- It is a system that is maintained/owned by the team.
- It is the central point for all team communication and meetings.
- It helps your team to understand:
  - what is needed
  - what the priorities are
  - who needs to do what and when
  - how the farm is performing.
- It makes the target vs actual condition clear.
- It drives efficient and effective decisions and actions.
- It encourages teamwork, ownership and accountability.
- It avoids the 'I didn't know' situation.

Lean Farm team activity

With your team, answer the following questions:

1. Discuss the farm's current visual boards (if any).
2. What information is currently tracked?
3. How visual and easy-to-understand is it?
4. What are some good things about the current boards?
5. What opportunities are there for using the visual management boards?
6. What actions can be taken to improve the current visual management board or to develop one?
7. What information should be on your team board?
8. Where is the best location for it?

That completes our discussion of the two key elements of visual management—visualisation and visual management boards. Now let's look at how visualisation should be applied to your farm's metrics to achieve your desired business results.

Your farm metrics

Many farms I have seen don't have very clear metrics beyond production and quality. Unfortunately, measuring only production and quality is not a holistic way of managing your business. For example, if the whole team is focused solely on how much you produce, other important elements of your farm such as maintenance, costs and animal health could be suffering as a result.

Metrics, measures, key performance indicators (KPIs)—or whatever term you wish to use—should be an essential part of your farming business.

Metrics help you ensure that:

- the business and team are focusing on the right things
- you are achieving your farm objectives and remain a sustainable business
- there is a clear target that the team can strive towards—a common goal
- you can measure immediately whether there is any deviation from the target and put in place immediate actions
- the team adheres to and sustains the right behaviours
- all aspects of your business are successful.

The team should most certainly be involved in setting up and agreeing to the targets. Importantly, once your farm has set some clear metrics and SMART (specific, measurable, achievable, relevant and time bound) targets around each metric, all the metrics should be made very transparent to the entire team. The team should be regularly monitoring and discussing each key metric so that everyone understands how they are performing against the metric—that is, whether they are above or below target. This means the team can take accountability for the metric and if performance is below target, the team can agree and take appropriate action to bring the metric back to target.

If you are a farm that doesn't have any or has very few metrics in place, then it is probably a good idea to start thinking about some and using at least a few key metrics that cover animals, production, quality, cost and safety as a minimum.

Examples of metrics on a farm include:

- *production* — litres or kilograms of milk solids (KgMS)
- *quality* — somatic cell count/grades
- *safety* — number of incidents/number of near misses reported
- *animal health* — body condition score, number of deaths, lameness, weights
- *people* — working hours, overtime, staff satisfaction
- *costs* — vet, maintenance, shed, feed
- *time* — milking time, calf feeding.

A Lean Farm example

Setting key metrics

A farmer mentioned to me that his empty rate was around twenty-one per cent. This appeared to be quite normal for him and he wasn't sure what he could do to actually improve it. Metrics are a fantastic way to turn around your farm's weaknesses, by placing more focus, measurement and monitoring on the weak-performing areas. If a particular element or process of your farm is not performing as optimally as it should be, then setting it as a metric with a clear target can significantly help to improve the element.

I suggested to the farmer that if this is a problem area for him, he should make it a key metric. Agree on a target that you want to achieve, for example 10 per cent. This means approximately a 50 per cent reduction in empty rate. The next step is to brainstorm with the team all the things you can do to reduce the empty rate. By tracking your empty rate against your target you

(continued)

A Lean Farm example *(cont'd)*

can easily see whether the actions you are taking are actually having an impact on the result.

You probably won't achieve your target in the first year, but with the right focus you may be able to over a couple of seasons. You can apply the same thinking to any other problem on the farm, whether it be maintenance costs, number of downer cows or number of hours worked.

Hopefully I have managed to convince you about the importance of having the right metrics. This next activity will help you start to establish some key metrics for your farm (if you don't already have some) by involving your team.

Lean Farm team activity

Agree on your farm metrics

1. Discuss the farm's current metrics, if any. What are they?

2. Brainstorm any other metrics the farm should use to achieve holistic success.

3. How are you performing currently against each of these metrics?

4. What is a SMART target to set for the season for each metric?

5. How will you capture, graph and monitor these metrics?

6. How often should you review and discuss each metric?

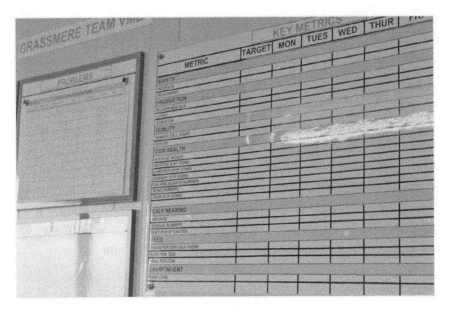

Our key farm metrics measured by the team

Visualise your metrics

There is no point in having metrics hidden away somewhere in a computer that no-one sees and only looking at them once a year. It is unlikely that your results will change very much this way. Your metrics have to be visual and regularly discussed so that your whole team knows them, understands them and can monitor how the farm is performing against each of them.

One of the greatest benefits a visual management board will have for your farm is if you use it to visualise your key farm metrics. The best way to visualise any metric is graphically rather than in a table. This is because a graph can very quickly tell you if you are above or below a target, what your trend is and if you have a problem. Anyone can glance at a simple graph and know what is going on.

Whether you are an owner-run farm, a one-person farm or have a large team, graphing your metrics, putting them up on a wall and monitoring them at least weekly will improve your ability to achieve your targets significantly. Just as in sport, if you have no target board how do you know if you are winning or losing? The same goes for business.

On our farm we have graphs for safety, quality, production, cost and animal health that our teams plot every week before the team meeting. This way the team own the data and are involved in monitoring the farm's performance. This is a very powerful approach.

In addition, the management visual board has a number of high-level graphs that are printed by the manager each week and discussed during the weekly management meeting (see figure 6.10 for an example). Again, the graphs help to very quickly visualise and assess the state of the business and identify any metrics that are below target or any poor trends. This means the business and team can identify poor-performing results immediately and respond to them quickly, taking action to turn around any negative results before it's too late to do anything about them.

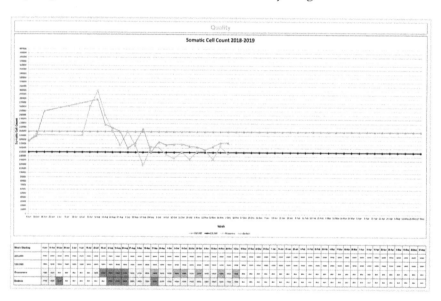

Figure 6.10: example of our quality graphs, which are printed and displayed on the visual management board and discussed weekly at the team and management meeting

Visual management team meetings

Once you have a visual management board set up on your farm, it should be the heart of all your team and management meetings. Each meeting should be run at the visual management board and the board should form the agenda for the meeting. The team should use the board to discuss and record all problems, activities, decisions, actions and plans in real time.

Mat reviewing the weekly metrics on the visual management board

These types of meetings are sometimes called 'stand up' meetings because the idea is that everyone stands around the visual management board during the meeting. The reason for this is to create a focused, interactive, efficient and concise meeting environment, where people aren't falling asleep in their seats, getting bored or starting multiple discussions—this is when meetings become a waste of time. This is still a work in progress in our team!

Rules for visual management team meetings include:

- it should be a stand up to keep everyone focused
- it should be short, sharp and to the point

- it should be a 5–30-minute planned daily or weekly meeting
- there should be a set standard day and time for the meeting
- it should be held around the visual management board
- no additional minutes should be recorded—the visual board is the minutes—and you can take a photo at the end for your records
- it should be an action- and performance-focused discussion
- it must be a two-way dialogue (not just pushing information down)
- detailed planning and discussions should be recorded as an action item and taken offline (to be discussed outside of the team meeting), otherwise the meeting will go well over time
- everyone owns the meeting and is accountable
- everyone is on time every time
- strong leadership is required: you must set expectations and support, lead by example and reinforce expected behaviours
- communication must be open, honest, constructive and sincere
- everyone should be respectful and open minded
- all members should challenge each other
- it should be transparent
- it must consist of teamwork.

Benefits of visual management team meetings

There are multiple benefits to visual management stand ups. I believe this is one of the key ways that any farm can truly transform its business. Benefits of a team meeting around the visual management board include:

- a consistent forum for communicating and sharing information
- the whole team understands the status of the business
- everyone is on the same page at the same time
- it is an opportunity to learn and develop each other
- the team can decide on and plan what actions are required immediately and prioritise
- it is a disciplined mechanism to drive the right actions and results

- it enables changes in behaviour over time
- it creates focus, ownership and accountability for the farm
- it achieves targets more effectively
- it builds an engaged and motivated team.

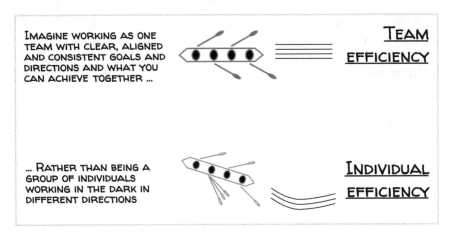

Visual Management team meetings drive a consistent focus and direction for the team.

Samii running the team meeting around the visual management board

What to discuss at visual management team meetings

There are no right or wrong answers in terms of what you should be talking about in your team meetings. It should depend on what is important for your farm, business and team and what information you have on your visual management board. Essentially, it is the content of your board that will guide all team discussions. Here are some guidelines as to what you could be discussing during your meetings based on what could be on your visual board.

Daily or weekly:

- Prior day's quality and performance issues
- Key activities for the day/week (e.g. AI, herd testing)
- Key actions for day/week
- Grazing plan
- Critical/hot issues
- Problems identified
- Problems and action sheet status
- Rewards, recognition
- Safety issues
- Maintenance issues
- Key metrics and trends
- Upcoming team meeting reminders
- 5S reminders
- Continuous improvement items

Less than weekly:

- Farm/industry news
- Changes in skills
- Training plans/courses
- Status of long-term process improvement issues
- Farm special projects
- Celebrations

Agenda for visual management team meetings

The brilliant thing about visual management boards and using them for your team meetings is that you don't need any special agendas. The board is essentially your agenda. This makes it very easy for anyone to simply follow the sequence of what's displayed on your board and you know that you have covered the essential things you need to discuss to run the farm effectively. Figure 6.11 is an example agenda for a weekly stand-up meeting.

WEEKLY TEAM MEETING AGENDA			
☑	TIME	ITEM	WHO
☐	2 min	Safety—any incidents/near misses/risks	All
☐	2 min	Key Communication—any important messages/information	All
☐	2 min	Problems—any new problems identified	All
☐	1 min	Maintenance—any new maintenance required/status update on existing maintenance	All
☐	1 min	Supplies—any supplies required	All
☐	1 min	Cows to watch—what needs monitoring	All
☐	3 min	Metrics—how are we going against target/graph trends	Herd Mgr
☐	1 min	People—how is team going/feeling	All
☐	3 min	Grazing Plan—what is planned for the week	Herd Mgr
☐	1 min	Season Plan—what's coming up	Herd Mgr
☐	2 min	Weekly Plan—what's happening this week/schedule	All
☐	3 min	Action Plan—what's everyone doing this week (what, who and when)	All
☐	1 min	Continuous Improvement—what's the focus for the week	All
☐	1 min	Other—any other issues/updates	All
☐	1 min	Celebration/Recognition	All

Figure 6.11: a Grassmere Team Meeting agenda

REMEMBER:
SHORT, SHARP, TO THE POINT — THAT'S WHY VISUAL IS KEY!

Lean Farm team activity

Work together with your team on the following actions:

1. Write down some of the agenda items that could be included in your team meeting.

2. Decide if you will have a daily or weekly meeting.

3. When will you have your meeting—what set day and time each week?

4. Identify what discussion points would be discussed daily/weekly or less frequently.

Summary

Visual management will . . .

- improve farm performance by identifying and reacting to problems in real time
- improve productivity by aligning people and actions more effectively and eliminating waste
- align everyone in the business to the farm's goals
- improve frontline leadership skills through better awareness
- improve morale through transparency, involvement and ensure that help/support is provided when needed.

VISUAL MANAGEMENT IS VITAL FOR ACHIEVING TARGETS!

Get your team to think!

A key component of visual management is to change the way your team thinks about their work and responsibilities. Visual management will help your team take ownership and start to think more and more for themselves rather than relying on others for information. It will give them more accountability and empowerment. This will have some big wins for your farm, team and business results.

This won't happen on its own or overnight, however. It needs the right leadership behaviours and support so that your team are empowered to think and take ownership. Here are some things you can do to help:

- Encourage people to think about things and come up with answers themselves before providing any answers.
- Ask questions—encourage challenging the status quo and having constructive discussions.
- Encourage new ideas and opinions.
- Urge the team to raise problems and see them as a positive.
- Check your ego at the door—your way may not be the only way or the best way.
- Genuinely admit that you don't know everything—no-one does and no-one expects you to and it's probably not in your job description either.
- Listen—you won't know any differently if you don't listen to what people have to say and to their ideas.
- Be willing to try new things and take sensible risks: testing things out will help you to improve much more quickly. It's an important part of continuous improvement.
- Always ask open-ended questions to stimulate people to think:
  - 'I don't know; what do you think about this?'
  - 'Why do we have a problem?'

- 'What is the problem?'
- 'What should we do?'
- 'What is the data telling us?'
- 'How can we do this/find out?'
- 'Why should we do that?'

REMEMBER: IT'S A SYSTEM NOT JUST A BOARD!

Lean Farm team quiz:
VISUAL MANAGEMENT

1. What are three examples of visualisation?

2. What does VMB stand for?

3. Who should own a visual management board?

4. What are the three rules of visual management?

5. What are three examples of information you could have on a VMB?

6. What is the best way to monitor metrics?

7. What is a visual management team meeting?

8. How often should you run team meetings?

9. How can you get people to think for themselves?

Lean Farm action plan:
Visual management

These actions are included to provide some simple guidance for your farm. They are aimed at giving you a little bit of motivation and direction if you need it. Of course, the idea is for you to do as much as you can to gain the maximum benefit of visual management on your farm.

1. Introduce visual management to your team.

2. Identify three areas of your farm that need visualisation (e.g. shed, office, paddock).

3. Take *before* photos.

4. Implement at least two forms of visualisation in each of the three farm areas.

5. With your team, design and mock up a visual management board.

6. Develop and install your final, marked visual management board in your team area or office.

7. Decide how the board will be used (the system)—e.g. frequency of meetings, who/how/when the information will be updated.

8. Start team meetings around your VMB.

9. Take *after* photos.

Now you are more familiar with how to find waste on your farm, how to create a well-organised, tidy workplace and how to use visualisation to improve communication. Your farm is well on its way to becoming a more Lean farm. The next tool that I will introduce, Standardisation, is considered the foundation of continuous improvement. Standardisation will help you sustain all the improvements and tools that we have discussed so far and underpins everything I talk about in this book.

Chapter 7

Standard work

Standardisation is the foundation of a Lean business. If we consider that Lean comes from the Toyota Production System (TPS), then Standardisation is the foundation of the famous TPS house. This is because it is the basis of continuous improvement (which is discussed in chapter 2).

So what is Standard work or standardisation? Here are some definitions to think about:

- *Standardisation* is the process of developing and agreeing upon a single set standard or method of doing a task or work activity.

- A *standard* is a document, instruction or accepted criteria or approach used by everyone doing a particular task or process. It establishes consistent practices within the business.

- *Standardised work* is an agreed set of work procedures that enable every employee to do a job in the safest, quickest, most reliable way with the best quality and lowest cost.

- To *standardise* is to choose the best method from many different ones and then use it repeatedly as the one way of doing that particular task or job.

EVERYONE DOING THE SAME THING   =   A STANDARD
THE SAME WAY, EVERY TIME

Lean Farm team activity

Draw a sheep

1. Ask each team member to draw a sheep on a piece of paper.

2. They have three minutes.

3. Display all the sheep on a wall for later.

Standardisation attempts to find the one best (current) way of completing a particular job, task, process or activity. It recognises that this standard or 'one best way' is not forever—there will always be better ways of doing the job in the future. However, it helps us to at least use the best approach that we know of right now.

Importantly, if you don't have any standard in place, it is very difficult to know whether there has been any improvement when something changes. You have no platform from which to measure any changes without a standard.

Standardisation should be owned and created by the people doing the work or job. This is the best way to ensure that you will develop the one most optimal approach and also gain buy-in from those who will need to follow the standard.

OUR FARMS ARE FULL OF REPETITIVE PROCESSES!

Standardisation is particularly useful when there are processes or tasks that need to be done regularly by different people. If it is just a one-off process or task that no-one will ever do again, there is probably not much point spending time trying to work out a standard.

However, if you have a process that needs to be done by a number of different people, or needs to be done more than once, then standardisation can be a significant benefit. Fortunately for farmers, a lot of what we do is repetitive or needs to be done by multiple people.

Our farms are full of processes and tasks that we need to do either every day, every week or every year. All of these tasks have a 'better way' and we can use standardisation to ensure that each of our tasks is done the best way possible, every time, by everyone, to ensure the best result.

Types of standards

Standardisation exists in all sorts of forms. You can standardise almost anything. Types of common standards include:

- standardised work (operating instructions)
- standardised meetings (set times and agendas)
- standardised routes/movement (the best way of getting from A to B)
- standards for safety (visual representation)
- standards for quality (accepted quality specifications)
- standards for 5S (what needs to be done, how, when and by whom)
- standards for maintenance (type and frequency of maintenance tasks)
- standards for animal wellbeing (how to care for and handle animals).

Standards come in a number of formats including check sheets, visual instructions, procedure manuals and e-learning modules.

Lean Farm team activity

Farm tasks

Think about all the tasks/activities you do on your farm.

1. Which ones are repetitive?

2. Which tasks or activities could you standardise on your farm?

Teat sealing—a job where standard work is important

Standardisation on the farm

There are numerous repetitive processes on our farms that could significantly benefit from standardisation. The list below covers a few of our most common farm processes. But standardisation can be applied even to the most basic of activities, such as putting up a fence reel, fencing, fixing water troughs, checking paddocks—all these tasks have one best way to get the job done that ensures the best efficiency and quality, and the least amount of overburden at the lowest cost and time.

There is nothing more frustrating than having 10 people do the same job 10 different ways with 10 different results. It not only causes a task to be inconsistent but also unreliable, not to mention frustrating. Having non-standardised processes will generate significant waste in your business and can have a big impact on results on your farm.

Examples where standardisation could be applied on your farm include:

- the milking process
- wash-down
- plant start-up
- moving cows

- establishing grazing plans
- pasture management
- maintenance of tractors
- cleaning of farm vehicles
- moving irrigators
- dry cow therapy
- mating processes
- yard cleaning
- animal treatment processes
- calving processes

- calf rearing/feeding
- calf shed preparation
- employee induction
- employee team meetings
- employee training and development
- quality checks
- cow marking
- effluent management
- a standard work week

Figures 7.1 to 7.5 on the pages that follow show examples of standardisation.

A Lean Farm example

Messy cow marking

Last year we were doing our teatsealing and a team of veterinary students came to assist. Our standard around what colour and how to mark the cows was not very clear, and therefore everyone ended up marking the cows in different ways. This caused a lot of confusion later and also potentially meant that some cows were missed.

If a simple task such as this isn't clear, it can generate a significant loss and cost to the business: team members wasting time asking each other what a mark means, missing marking altogether, not recognising a mark or even seeing it. This causes confusion, frustration and potential mistakes.

As a result we have now developed a clear visual standard for all our markings so that everyone does it the same way and everyone understands what a mark means. We also use a one-page standard document to train any new team members or external people who assist us with the process.

KEY POINT LESSON: Daily Mating Process

1. SORT COWS TO MATE/RECORD
2. CHECK SENSORS/EQUIPMENT
3. CHECK AFI DRAFTING OK
4. DRAFT COWS FOR AI
5. OBSERVE COWS ON FEEDPAD
6. IDENTIFY FALSE POSITIVES
7. CHECK BOOK/AFI LIST
8. PUT HEAT COWS ON PLATFORM
9. RECORD IN BOOK IN ORDER
10. COWS READY FOR AI BY 8AM
11. HELP AI TECH/ ASK HIS OPINION
12. GET COWS OFF PLATFORM
13. CONTACT BODMIN NEAR END
14. COWS SPLIT IN RIGHT HERD
15. FINAL SHED HOSE/CLEAN

Grassmere Dairy Mating Key Point Lesson v1.0

Grassmere dairy

Figure 7.1: our 'Keypoint Lesson' standard document for mating

MY STANDARD WEEK: SHED ROLE GRASSMERE

TIME	MONDAY	TUESDAY	WEDNESDAY	THURSDAY	FRIDAY	SATURDAY	SUNDAY
4:30-5	Start up shed	Start up shed	Start up shed	Start up shed	Start up shed	Start up shed	Start up shed
5-5:30	Milking	Milking	Milking	Milking	Milking	Milking	Milking
5:30-6	Milking	Milking	Milking	Milking	Milking	Milking	Milking
6-6:30	Milking	Milking	Milking	Milking	Milking	Milking	Milking
6:30-7	Milking	Milking	Milking	Milking	Milking	Milking	Milking
7-7:30	Milking	Milking	Milking	Milking	Milking	Milking	Milking
7:30-8	Shed clean up	Shed clean up	Shed clean up	Shed clean up	Shed clean up	Shed clean up	Shed clean up
8-8:30	Yard clean	Weekly Team Meeting	Yard clean	Yard clean	Yard clean	Yard clean	Yard clean
	Daily Stand Up 5 min		Daily Stand Up 5 min	Daily Stand Up 5 min	Daily Stand Up 5 min		
8:30-9	Breakfast	Breakfast	Breakfast	Breakfast	Breakfast		
9-9:30	AFI Data/ Maintenance/ Records	Yard clean	Weekly CI Session	AFI Data/ Maintenance/ Records	AFI Data/ Maintenance/ Records		
9:30-10	AFI Data/ Maintenance/ Records	AFI Data/ Maintenance/ Records	Weekly CI Session	AFI Data/ Maintenance/ Records	AFI Data/ Maintenance/ Records		
10-10:30	Pedometer sorting/ checking	Pedometer sorting/ checking	AFI Data/ Maintenance/ Records	Pedometer sorting/ checking	Pedometer sorting/ checking		
10:30-11	Pedometer sorting/ checking	Pedometer sorting/ checking	Pedometer sorting/ checking	Pedometer sorting/ checking	Pedometer sorting/ checking		
11-11:30	Internal 5S	Internal 5S	Internal 5S	Internal 5S	Internal 5S		
11:30-12	Shed Maintenance – strings	Shed Maintenance – strings	Internal 5S	Shed Supplies	Shed Maintenance – strings		
12-12:30	LUNCH	LUNCH	LUNCH	LUNCH	LUNCH		
12:30-1	LUNCH	External 5S	LUNCH	LUNCH	LUNCH		
1-1:30	External 5S	External 5S	Shed Maintenance – strings	Shed Maintenance – strings	External 5S		
1:30-2	VAT wash/ shed start up	VAT wash/ shed start up	VAT wash/ shed start up	VAT wash/ shed start up	VAT wash/ shed start up	VAT wash/ shed start up	VAT wash/ shed start up
2-2:30	Milking	Milking	Milking	Milking	Milking	Milking	Milking
2:30-3	Milking	Milking	Milking	Milking	Milking	Milking	Milking
3-3:30	Milking	Milking	Milking	Milking	Milking	Milking	Milking
3:30-4	Milking	Milking	Milking	Milking	Milking	Milking	Milking
4-4:30	Shed clean	Shed clean	Shed clean	Shed clean	Shed clean	Shed clean	Shed clean
4:30-5	Yard clean	Yard clean	Yard clean	Yard clean	Yard clean	Yard clean	Yard clean
5-5:30							
5:30-6							

Figure 7.2: Grassmere Shed role standard work week

Cleaning Process on Monday & Thursday!!

A After Step **33** Fill Jetter tub with hot water at 80C (500L)

B Add 500gms of Klear Klenz to tub

C 2 x week add 500mls of XY-12 if required

D Recycle through plant for seven (7) minutes and dump down drain at 60C min

E Fill Jetter tub with Cold water (500 litres)

F Add 500 mls of Acquaduz (Acid)

G Rinse plant system out and dump remaining water

H Continue with step **42** – final cold rinse with 250 L

DAIRY PLANT CLEANING SCHDULE!

EVERY MORNING AND EVENING AFTER MILKING:

1. FLUSH OUT THE PLANT WITH 500+ LITRES OF COLD WATER
2. PRE-HEAT PLANT WITH 80 LITRES OF HOT WATER (80C)
3. WASH PLANT WITH 500 LITRES OF HOT WATER (80C) WITH 750MLS OF **ALLTEMP**
4. RECYLCE FOR SEVEN MINUTES AND DUMP AT 60 C MINIMUM
5. DO COLD WATER RINSE WITH 250 LITRES

EVERY MONDAY AND THURSDAY

1. FLUSH OUT THE PLANT WITH 500+ LITRES OF COLD WATER
2. PRE-HEAT PLANT WITH 80 LITRES OF HOT WATER (80C)
3. WASH PLANT WITH 500 LITRES OF HOT WATER (80C) WITH 750GMS OF **KLEAR KLENZ**
4. ADD 500 MLS OF **XY12** ONCE A WEEK IF REQUIRED
5. RECYLCE FOR SEVEN (7) MINUTES AND DUMP AT 60C MINIMUM
6. RINSE OUT WITH 500 LITRES OF COLD WATER WITH 750MLS OF **ALLTEMP**
7. DO COLD WATER RINSE WITH 250 LITRES

3 | Grassmere Dairy Shed Plant Cleaning Process v1.1

Grassmere dairy

Figure 7.3: one of our standard work documents for cleaning the plant after milking

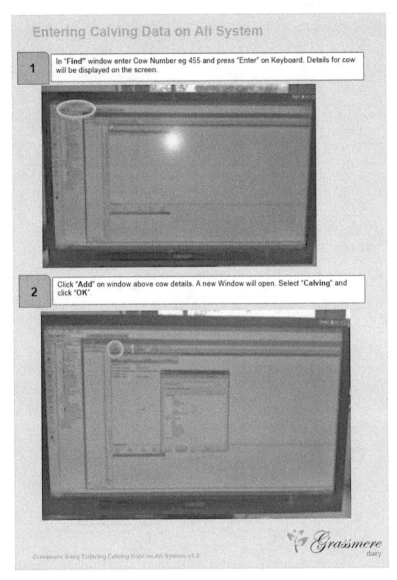

Figure 7.4: Grassmere standard process for calving data entry

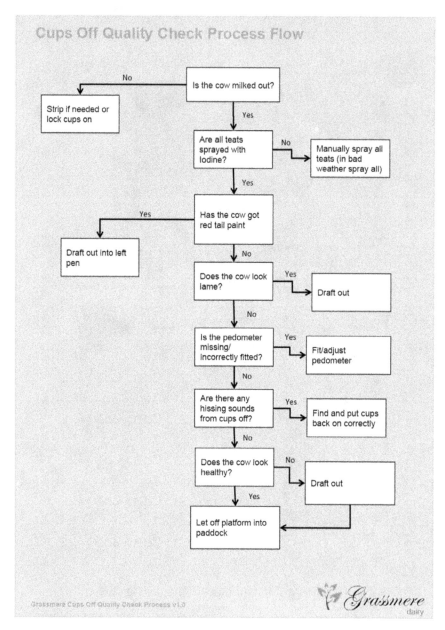

Figure 7.5: quality check flow diagram

Lean Farm team activity

Walk around your farm and think about all the things you do every day.

1. Observe the current situation regarding standardisation of work.

2. What examples of good standardisation can you see on your farm? (Take photos.)

3. What opportunities exist for standardisation on your farm? (Take photos.)

4. What actions can be taken to improve the level of standardisation on your farm?

Benefits of standardisation

Not only is standardisation the foundation for continuous improvement, but it can have an enormous effect on building a highly productive, efficient, safe, sustainable and successful farm.

Farmers who don't have standardisation in place can regularly feel like they have no control, inconsistent quality, very little discipline and large variation. This results in constant frustration, long work hours and stress. Once standardisation is adopted, you will have better control, less variation, improved quality, less waste and improved team satisfaction. Importantly, it will also improve your productivity and reduce time spent on jobs.

Standardisation also helps set clear expectations. Without standardisation it is very difficult to make people accountable for errors and poor quality of work. If someone doesn't know what is expected of them because there are no standards in place, it is unfair to blame them when something is not done as was expected. Having a clear standard in place for a task ensures that everyone is very clear on what is expected and they can then be held accountable.

Standardisation is beneficial because:

- it eliminates waste
- it sets clear expectations
- it improves the consistency and reliability of a job
- it can be measured
- it improves quality
- it reduces the time needed to complete a job
- it reduces cost
- it improves safety
- it improves customer satisfaction
- it reduces stress and confusion
- it improves team work and culture
- problems can be identified and addressed more easily
- it makes induction and training of team members easier.

A Lean Farm example

What's the right way?

A farm manager told me a story about a situation he had experienced on his farm in relation to a paddock gate that needed attention. He had asked one of his team to fix the gate before the cows went into the paddock. The team member fixed the gate and that afternoon when the cows were moved to the paddock they managed to break out of it. The manager went to see what happened and noticed immediately that the gate had not been fixed properly, enabling the cows to easily break through. The manager ended up doing rework, fixing someone else's poor quality of work and wasting two hours of his time. When the manager addressed it with the team member, he was told that it had been fixed well enough. If there is no standard of what good is, then it is hard to argue that it wasn't done 'well enough'.

Potential saving: two hours of time

This is a perfect example of what happens on our farms every day...and it *wastes* our *time*. We all assume that even simple tasks like this are common sense and everyone knows how to do them. However, even if everyone is able to do it they probably won't do it to the level you expect. This is because each of us has a different idea of what is considered 'acceptable' or a 'quality job'. What is considered 'acceptable' or 'good enough' to me may not be 'acceptable' to you. Unless this expectation is written down and everyone knows about it, it is very difficult to argue with someone that the way they completed a task was wrong.

The statement that follows makes a very important point. I often hear farmers complaining about how someone has done a job incorrectly and created some problem. But what's to say that your method is the right method? It is very difficult to argue with someone that their way is not right and they should follow your method, particularly if there is no evidence that your way is the best way to do a job.

Standardisation creates one 'right way'

If, however, you spend time with the whole team analysing a particular process, identifying the best known method at this point in time, agreeing to it and then developing a standard of work describing this best method, you are in a much better position to know what is the 'right way'. It is then much easier to have a discussion with someone when the agreed standard or 'best way' hasn't been followed.

Flexibility and standardisation

Often people believe that if they standardise they will lose flexibility. This is a big misconception. Standardisation most certainly does not restrict flexibility on your farm. Standardisation makes your jobs more efficient but also helps you to more easily spot all wastes and therefore improve your processes more frequently. It improves a business's flexibility as it helps more people to do the same job in the right way. It also helps a business to be much quicker to adapt to change, as the business knows exactly how it is doing a particular job and therefore can make changes more easily and in a more controlled way.

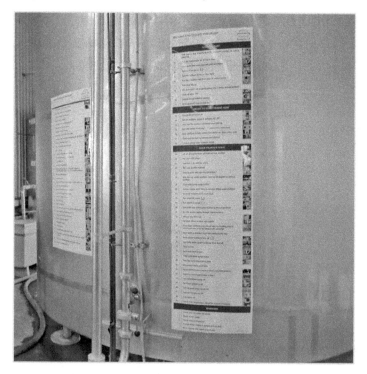

Example of standardisation: a standard checklist for the milking procedure
Source: Courtesy of VizLink

A Lean Farm example

It's all about your method

Have you ever experienced one person milking the whole herd in an hour while the rest of the team take two hours? This can be a result of poor standardisation. It doesn't mean that everyone should be able to do it in one hour (you need to have a process that is realistic, achievable and sustainable by the majority of people). What it does mean, though, is that there is opportunity to improve. By analysing the best approach to doing something you can work out the one best way and agree on a standard to reflect this that everyone follows.

A farmer I was talking to experienced this situation.

By observing the fast team member as well as the others, the team could very quickly spot the differences in approach. These included walking distance, sequence of putting on cups, how the person moved and general animal handling. The team could then identify a method that would be achievable and improve everyone's milking time. Improvements included reducing the walking distance by introducing a marked floor space within which team members should remain. The farm was able to reduce the milking time by around 45 minutes per milking.

Saving: 45 min of milking time × 2 people × 2 milkings

= 3 labour hours/day

= 1095 hours/year

= 109 labour days/year (based on a 10-hour work day)

(continued)

A Lean Farm example *(cont'd)*

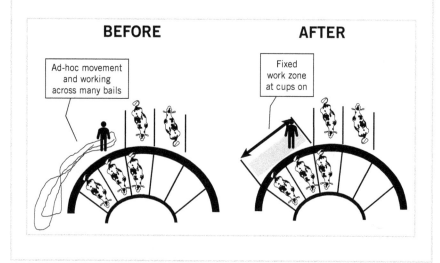

A farm without standardisation

Most farms have very little standardisation other than the processes that are subject to external regulations, such as quality parameters. In today's world, our farms need to change frequently to keep up with changing consumer and society demands, environmental changes, regulations, technology and the economic landscape, including markets. To adapt and comply with these changes, farms need to adopt new processes or change and improve existing processes.

If we don't have any standardised processes in place, and all our work is done in ad-hoc, inconsistent ways, then how can we:

- know where the problems are?
- know what we want to improve?
- know how to improve it?
- explain to our team what and how we want to change if everyone is doing something different?

Furthermore, once we have made a change, if we didn't have any standards in place originally, how will we:

- know if the change has made an improvement to the process or task?
- know if the change has made it worse?
- revert to the original process or condition if the change doesn't work?

Typical features of a farm with no standards include:

- significant rework
- always fixing things that weren't done right the first time
- people waiting around for things, machines and others
- too much walking around
- driving back and forth
- poor meeting discipline: people are late/no shows or meetings take too long
- repetitive mistakes/problems
- constant management required
- constant firefighting, particularly by management
- low team morale
- frustration and stress
- accidents
- milk quality grades
- animal deaths from incorrect diagnosis/treatment
- a high somatic cell count.

A Lean Farm example

Calf chaos

I was at a farm during calving watching three team members try to sort out calves in the calf pens. One member was trying to match collars with some scribbled-down notes in a crumpled-up pad; the other was lifting calves from one pen to another. Then the one scribbling notes said something and calves were transferred back again. The third guy was walking around scratching his head looking very confused. There were calves running all over the place. Different numbers in different pens. Some with no collars or any identification at all. It simply looked like chaos and I am pretty certain it was. There most certainly wasn't any standardised work! The result: most probably mixed-up calves, wasted time and cost.

Calving and calf management is a process you want to get right. In this case standardisation could have been applied to cow/calf identification, calf data recording, calf pen set up and management, and calf handling.

Don't blame the person

It is so easy to blame someone when something goes wrong. Whether it is the trough not being fixed correctly, a downer cow not being diagnosed correctly, calves being put in the wrong pen, the motorbike oil filters being choked due to lack of oil, handles on fences getting placed in the wrong spot so the fence isn't hot or cows breaking out, we are very quick at blaming. However, while it may be a genuine mistake, these problems may also be the result of a lack of processes and systems.

Whenever something goes wrong, rather than being quick to blame someone, always ask the following questions, in this order.

1. *Is there a standard for this?* Does your farm have a standard in place for the task that went wrong? This doesn't mean that it is in someone's head but rather that the standard has been documented in some way.

2. *Is the standard easy to understand?* Has it been clearly documented? Will anyone understand what it means if they see it? There is no point having some 1000-word standard procedure in place if no-one understands what it means. A standard is only effective if it is crystal clear to everyone how the process or task should be done.

3. *Does the team member know about the standard?* Yes, you might be thinking this is ridiculous but it is surprising how many times I have seen a situation where standards exist and have been around for a long time but are either hidden somewhere on a computer or in a filing drawer and no-one has ever seen them or heard about them. Not a very useful standard, is it? Once a standard has been developed, it must be visualised, communicated regularly and accessible when needed by the team.

4. *Is the standard process being followed?* If the answer is 'no' and answers to all the above questions are 'yes', you need to have the right discussions with the individual. Very rarely you might also get a 'yes' response to this question. If this happens, you need to review what happened and why, whether your standard is correct and truly clear and do what's necessary to ensure that the problem never occurs again.

From my experience, the breakdown most often happens at one of the first three questions. It is very rare that you can honestly answer yes to the first three questions and arrive at question 4.

A simple mistake is not understanding when a tape gate is hot

Never assume

There are many jobs on the farm that seem very straightforward and common sense to the seasoned farmer. However, just because something seems like the most basic job under the sun to you, it doesn't mean that every person in your team will know how to do that job.

From what I have seen, farmers generally aren't the most gifted teachers. This is because many farmers have been in the game so long that they expect others to just know things, or it genuinely doesn't even occur to them that someone may not know how to do what seems to be a very simple task. Furthermore, many farmers understandably don't have a degree in HR and so are probably not familiar with training techniques to help develop their team. When asking someone to go and do something, a farmer will generally say something like 'Can you go and set up that break?'—while racing off to fix some other problem—without any further explanation and expect the break to be done correctly.

I have heard of farm employees who have been working on farms for several years yet no-one has ever shown them how to do some of the most basic tasks effectively. They have just worked it out somehow on the go. This means that, potentially, even employees with 'experience' may not actually know how to perform common tasks correctly. What's more, if they were taught how a particular job is done, the method they were shown once upon a time may not be the best approach to use on your farm. We have also seen examples on our farm where team members who have spent years milking cows, for example, didn't actually understand how the milking process really worked as no-one had ever explained it to them. It was just assumed that everyone knows how a milking plant works.

So don't ever assume people know what to do. Try to regularly do 'refresher' courses with your whole team about all processes and equipment or machines so that no-one feels uncomfortable or embarrassed. Never single someone out in front of others or belittle them for not knowing how to do a simple task. This will just ruin your team culture.

A Lean Farm example

Well ... no-one actually ever taught me how to ...

One farmer shared a couple of funny examples with me of farm employees not understanding how to do what appeared to be basic tasks.

The farmer asked one of his team members to put up some waratah standards and gave him a rammer. A little while later, the farmer heard 'dong', 'dong', 'dong' and went to see what was going on. The team member was working hard trying to put the standards into the ground. However, he was using the rammer like a hammer and trying to hammer the standards in!

In another example, a team member had the waratah standards pointing sharp end up and was trying to hammer the flat end into the ground!

Needless to say, *never assume anything!*

Standardisation = continuous improvement

Standardisation is the foundation of a Lean company. It is the process of standardisation that enables us to analyse any process or task that is part of our job, identify all the waste in that task, eliminate the waste and develop a better process for doing the task. This is continuous improvement, as discussed in chapter 2. Through the process of standardisation we are able to improve our overall productivity, including the time it takes to do a job, the quality, the cost, the safety and the ease of the job.

Standardisation helps us improve the way we do things in a controlled manner. If we have a very clear standard in place, we know exactly where we started from and can easily go back to that process. Importantly, we can also measure the benefit any change has created by directly comparing the previous process with the new process. This means we know if we have tangibly improved our farming process.

STANDARDISATION IS NOT SET IN STONE.

Remember, once you have standardised something it doesn't mean it is set in stone. The particular task should continue to be improved as better ways of doing the job are identified. By actively reviewing your existing standard processes, you can scrutinise them further and continue to refine and improve them. Once improvements to the process are identified, the standardised work is modified to reflect the changes in the process and to prevent it from sliding back to the previous method.

The PDCA model

The PDCA (Plan, Do, Check, Act) model is a Lean tool used to improve processes (it is also known as the Deming cycle). This loop is continuous. Every time you make an improvement, you update the standard to stop the process slipping backwards to the previous process. Standardised work is like the wedge that is preventing the improvement from rolling backwards. See figure 7.6 (overleaf).

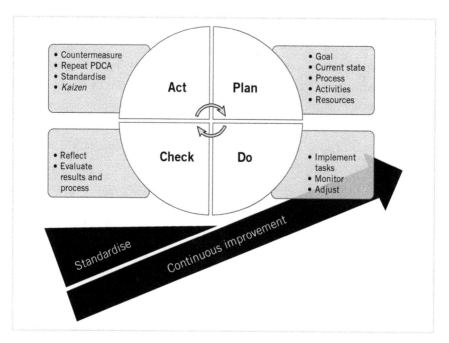

Figure 7.6: the PDCA model

Lean Farm team activity

Draw a sheep (2)

Review with your team the pictures of sheep that they drew earlier in this chapter.

Then show them this picture and tell them that this is how the sheep should look.

1. Ask each team member to draw the sheep in the picture.

2. They have three minutes.

3. Then ask them to hold up their sheep.

4. Discuss the following with your team:

   — What do the sheep look like now compared to the first ones?

   — Do they look more similar to each other?

   — Do they look closer to what the customer had wanted than the first lot?

   — Why do they all look very similar?

   — Was it easier to understand what was expected this time? Why?

   — Can you think of a farm example where a simple visual standard like this will help you?

How to develop a standardised process

Dictating to someone what to do and how to do it will not ensure people believe in the process and stick to it. They also need to understand *why*. If a person is involved in developing a process, they are more likely to understand why a job should be done a particular way and why this is the best way. They also have to agree to follow the process. This will significantly improve the chances of the process being adopted and sustained by all team members.

There is no right or wrong way of developing a standard process. Placing some photos showing the key process steps on a page is a good start. There are, however, various standard approaches that have been developed and are used by the likes of Toyota. The structured approach that I will describe to you is particularly useful for more complex processes that are done in different ways by different people and are likely to cause disagreement.

As I mentioned, it's important that you involve your team in the development of the process. Start by looking at the existing process, videoing it and then analysing it with your team. This way everyone can watch the process impartially and there is no feeling of being under the microscope. It also means you can replay the video multiple times to really spot all the opportunities for improvement.

1   Video the existing process

Videoing a process is the easiest way to visualise and analyse it.

a   *Video.* Before videoing a process, ensure the whole team is familiar with it. Then get the team to do the videoing (see figure 7.7). Start with the simpler regular processes that have a clear starting and finishing point. As you become more experienced in developing standardised work, focus on more complex processes.

b   *Watch the video and record the steps.* Ensure the whole team watches the video together. Record all the steps that you observe in the process on a flipchart or whiteboard in sequence. Ensure you have captured all the work detail and key steps including any walking, motion or waiting.

c   *Allow for multiple people doing the same process.* If there are jobs that several team members take turns doing and each person does the job differently, you might want to take multiple videos—one for each method. This way you can easily analyse each approach. (See step 3 for ideas about selecting the best method.)

Figure 7.7: videoing our feedout process

2   Analyse the process

This step is critical. To get a great standard process, this step needs to be done well. Importantly, this step needs to be done with the full team to ensure engagement.

a   *Review the current process.* Read through all the steps you wrote down in step 1 and watch the video again.

b   *Identify the value-added steps.* Get the team to identify all the components of the process that appear to be value added (remember value add is in the eyes of the customer). Highlight these steps in green.

c   *Identify the non–value added or waste steps.* Ask the team to scrutinise the steps and decide which steps are pure waste. Consider the 8 wastes and try to identify any of these in the process. Highlight all the waste steps in red. This may be difficult, so the team need to humour each other and be as honest as possible with each other.

d   *Identify all current concerns.* Think about any other problems you are currently having with the process. This could be poor quality, inaccuracy, accidents, maintenance issues or high costs. Write these down on a separate flipchart. Ideally the new standard work should address most, if not all, of these issues.

e   *Identify opportunities for improvement.* Finally, ask the team to spot any other opportunities for improvement in the process. You can do a general brainstorm to capture any ideas that may not have been identified or discussed already.

3   Agree on the best method

If there are various methods for doing the same job, now is the time to determine with the team which method is best. The best method may be a combination of one or more existing methods.

a   *Review the processes and opportunities.* With the team, review all the processes again and the waste and opportunities identified in each process.

b   *Select the best approach.* Agree on which process steps should remain and the best method to use. While considering all the observations made during step 2, you can also use a matrix like the one shown in table 7.1 to assess each method in terms of various criteria such as ease, quality, safety, time, cost, repeatability and sustainability. This can help you to quickly assess which method is the most appropriate.

Table 7.1: method assessment matrix

Method	Safety	Ease	Time	Quality	Cost	Repeatability
A	✗	✓	✓	✓	✓	✗
B	✓	✓			✓	✓
C	✓	✓	✓	✓	✓	✓
D	✓	✗	✓	✓	✓	✗

4   Remove waste steps from the process

The team should watch the process video made in step 1 again and focus on all the steps that fall into the 'waste' category (for a reminder about the 8 wastes see chapter 4).

a   *Analyse each waste step.* Each step should be analysed and the team
    should brainstorm ways of removing each one. To do this you can
    ask the following questions:

    •   What is the purpose of the step?

    •   Why does it need to be done?

    •   Can it be eliminated?

    •   Can it be done another way?

    •   Is there a way to reduce or improve the step?

b   *Agree on an action plan to improve.* The team should agree on the
    specific actions to take to eliminate the waste (or at least improve it
    by reducing time taken to do the step).

5   Balance the work

If the process that you are analysing is one that requires more than one
person, it is helpful to analyse the activities that each person is doing. This
way you can quickly see the amount of work each person has and try to
balance the activities evenly so that the total process time is the same.

a   *Visualise the tasks involved on a time scale.* You can create a chart
    with a time scale in minutes or hours and allocate each step of
    the process on the graph as a time increment. You can draw—or,
    preferably, use sticky notes—to do this. This way you will be able
    to easily see if one person has a higher workload than another.
    If this is the case, the total process time will potentially be
    unnecessarily long for one person, while the second person may
    be waiting (one of the 8 wastes discussed in chapter 4) for the other
    person to complete their tasks.

b   *Re-distribute work tasks.* Try to identify tasks that can be
    re-allocated to the person with the lower workload. This may
    require changing the work sequence. Re-allocate tasks as needed
    to make the total work time the same for each team member.
    Figure 7.8 (overleaf) illustrates this.

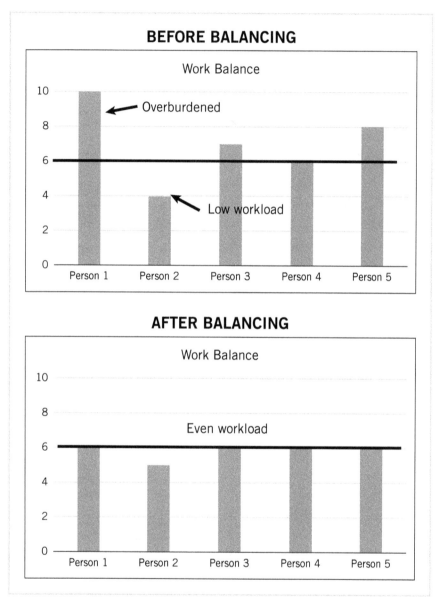

Figure 7.8: workload before and after balancing

6 Develop the draft standard document

It is time now to document the standardised process for the particular job. If the first five steps have been done well, this step should be straightforward.

a   *Finalise the process.* Sketch out the final agreed steps in the process. If there are multiple people involved in the process ensure that the steps for each person are defined. Visualise the order of these steps on a wall or whiteboard so that the whole team can see them and imagine them.

b   *Decide how to best standardise the process.* Should you use a visual document, a one-page document, a checklist? What will the standard look like? Sketch the standard out on butchers paper so that everyone can understand it and agree to it. Ensure the whole team have agreed to the final standard.

c   *Document the process.* Allocate one or two people in the team to document the agreed process into a formal document template. The document template should be version controlled and clearly titled. It should also be in a very easy, visual format. You can ask the whole team to sign off on the document to show their approval and commitment.

Lean Farm team activity

Develop a standard work document

1. As a team, identify any problems/mistakes that have happened on your farm recently.

2. Could a clear standard process have prevented these problems/mistakes?

3. Discuss the farm process that would benefit from being standardised.

4. Develop a standard work document of your choice for the process.

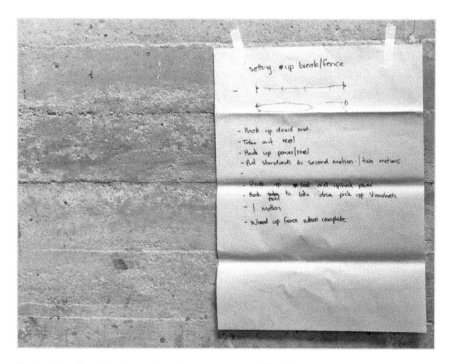

Brainstorming the key steps for a process with the team

7   *Test the new standard process*

Once the standard is in place you need to test it to ensure that it will actually work. Run a trial of the new process and capture any issues identified with it. Check that the process has in fact produced the expected improvements. It is unlikely that the process will be perfect the first time—that's okay! Continuous improvement is all about constant trialling, adjusting and trialling again until you get closer to perfection.

8   *Improve*

During your trial, the team may have identified some issues with the newly developed process. You may need to make some tweaks or improvements to the process to get it right.

   a   With the team, identify solutions to the concerns found at this stage.

   b   Implement the solutions and test the process again.

c   If the solution or improvement works and solves the problem, you
can update the draft standardised work document.

9   Finalise the standard document and make it available

The standardised work document that you have created can now be
finalised with the iterated process. Make all changes necessary and sign off
on the document.

Make it available in an appropriate location that can be easily accessed by
all team members.

10   Train your team

Now that you have a final process agreed on and documented, it is vital
that everyone is trained in the new method so they all understand it and
can action it. Using a skills matrix (in a similar format to the method
assessment matrix in table 7.1), for example, showing all your team
members' names against processes they are trained in, can help to track
everyone's knowledge and skill level against each process or standard.

11   Seek continuous improvement

Voilà! You should now have an agreed standard for a particular process
or job. However, that doesn't mean this is the process that will be used
forever.

a   *Improve.* You need to ensure that you regularly review all of your
processes and improve on them. If someone identifies an even better
way of doing a particular task, the process should be updated to reflect
the improved method. Remember that standards are not set in stone.

b   *Manage change.* You also need to be realistic about updating
standards. You should not change your work standard every day!
This will just confuse everyone and no-one will know which
standard is the latest.

Having a standard approach to managing standard work changes will
enable you to make the required updates in a controlled, sustainable way.

This could involve collecting all the ideas for improving a particular process and, every three or six months, making a bulk change to the process, incorporating all the individual ideas. Of course, if there is a critical improvement it should be made as soon as possible.

The best standard document

The best type of standard document is simple, short and to the point. Ideally it should be one page long and:

- key information should be easy to identify and read
- be visual (pictures, drawings and diagrams)
- use colour to make important elements stand out
- be located at the point of use where the job is done, so that the team can refer to it as needed
- be used as the standard for training team members
- be *dated*, state who the *owner* is, if relevant, and be *version controlled*.

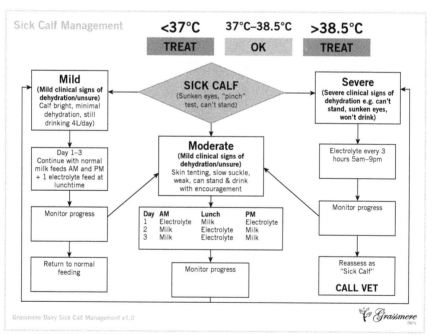

Our standard process chart for sick calf management

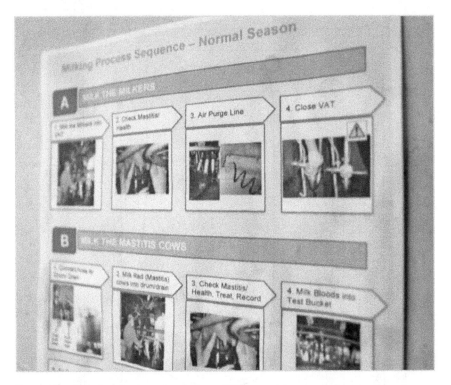

Example of a standard process document used on our farm

The standardised work chart shown in figure 7.9 (overleaf) is a common Lean-style template that you can use to document your standard processes. It is simple and contains all the information on one page including the key steps, any photos, diagrams of the process and a layout of how the work or person should flow/move to complete the task.

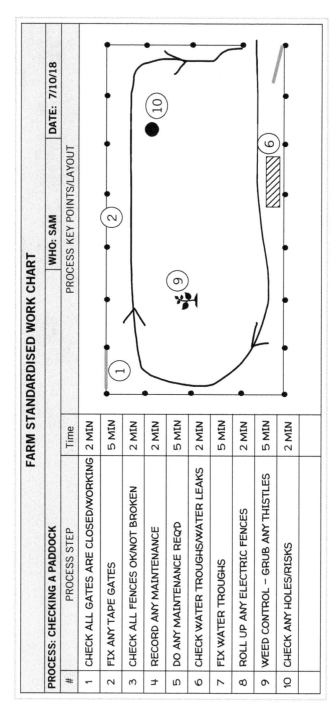

FARM STANDARDISED WORK CHART

PROCESS: CHECKING A PADDOCK		WHO: SAM	DATE: 7/10/18

#	PROCESS STEP	Time	PROCESS KEY POINTS/LAYOUT
1	CHECK ALL GATES ARE CLOSED/WORKING	2 MIN	
2	FIX ANY TAPE GATES	5 MIN	
3	CHECK ALL FENCES OK/NOT BROKEN	2 MIN	
4	RECORD ANY MAINTENANCE	2 MIN	
5	DO ANY MAINTENANCE REQ'D	5 MIN	
6	CHECK WATER TROUGHS/WATER LEAKS	2 MIN	
7	FIX WATER TROUGHS	5 MIN	
8	ROLL UP ANY ELECTRIC FENCES	2 MIN	
9	WEED CONTROL – GRUB ANY THISTLES	5 MIN	
10	CHECK ANY HOLES/RISKS	2 MIN	

Figure 7.9: standardised Work chart for checking paddocks

How to sustain standard work

It is easy to set up a heap of standard work documents, but they are not very helpful if no-one follows them. Sustaining the standard way of working can be very difficult. Even if the team is highly engaged and follows the agreed standard initially, over time they may slip into bad habits again.

Obviously, if you start with an enthusiastic and engaged team you will be in a much better position to sustain your standards. Here are some tips to help you sustain your standards as much as possible:

- Ensure your developed standard is based on facts and analysis, not on assumptions and opinion. This makes it objective, not subjective.

- Make sure everyone understands the importance of standardised work and why they need it.

- Ensure it has been developed, discussed and agreed on by the whole team. Without this buy-in it will be very difficult to introduce any standardisation.

- Reinforce that the process belongs to the team—they are the masters at the job and need to own it and continue to improve it.

- Regularly observe, measure and assess the standard process.

- Visually document and display the standard at the point of use as a constant reminder to everyone.

- Lead by example, and constantly refer back to the standards and reinforce their importance. Never go past a non-conforming situation and ignore it—address it immediately.

- Hold everyone responsible for maintaining and following the standard.

- Team leaders or managers should be held responsible for ensuring compliance by the team.

- Challenge the standard regularly so that further improvements can be made.

- Retrain team members regularly in the standard to prevent bad habits from developing.

- Audit the process regularly to check that it is still working and being used.

While empowering your team to take ownership of the processes and be involved in developing their own standard work is the ideal approach, it most certainly won't be the quickest. If you need a process in place immediately, you may need to draft something with just a couple of people involved and then evolve it over time with the rest of the team, gaining everyone's input. Of course, you need to communicate this to your team so that everyone is aware of what you are doing and why you are doing it in this way.

Audits

Audits are necessary to regularly check in with the process and ensure that the job is still being done according to standard and there is no deviation. Audits should be done at different levels—for example, owner/sharemilker, manager and even team member (e.g. farm assistant)—see figure 7.10. Successful audit systems are also standardised! Audits should:

- include a robust standard tiered audit system
- be carried out at a standard frequency
- be organised so that both the standard work documentation and the physical job are audited
- identify any discrepancies immediately and these should be recorded on the audit sheet and discussed with the individual team member and if relevant the team
- tailored to suit your processes.

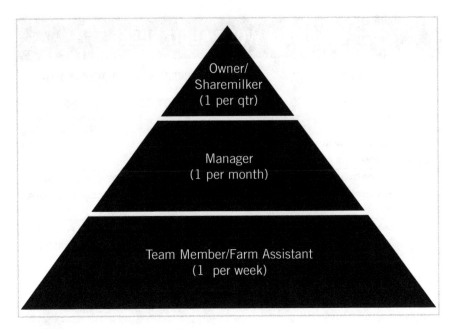

Figure 7.10: a tiered audit system

Standard work audits can be combined with 5S/visual management (see chapters 5 and 6) and other audits such as a monthly general farm audit. This makes the process of doing them more streamlined and efficient, and easier to remember. See figure 7.11 (overleaf) for a standard work audit sheet.

STANDARD WORK AUDIT SHEET

PROCESS: CUPS ON		WHO CHECKED: JACK	DATE: 04/01/18

AUDIT CHECK ITEM	OK	NOT OK	COMMENTS
DOCUMENTATION			
Is there a standard work document for this process?	✓		
Is the standard work document visual and displayed at point of use?		✓	Not very visual and displayed far away
Is the standard accurate/valid?		✓	No needs a point about new display screens
Is the document easy to follow/understand?		✓	Too wordy
WORKPLACE			
Is all PPE available as per standard?	✓		
Does workplace layout match the standard?	✓		
Are all items positioned as per standard?		✓	The paint cans aren't there
Is standard suitable for the workplace environment/layout?	✓		
PEOPLE			
Does team member know about the standard?	✓		
Does team member understand the standard?	✓		
Does team member use/refer to the standard?	✓		
Is team member following the standard?		✓	Not using correct sequence of applying cups
Can team member identify any opportunities for improvement?	✓		Need better labels on control panel and markings on floor

Figure 7.11: an example of a standard work audit sheet

Summary

Standardised work is vital for a well-run, organised and efficient farm. It is the foundation for continuous improvement and without it validating improvements can be difficult.

Standardised work can have a significant impact on your farm. It will help your farm to:

- be more productive and effective
- be compliant and safe
- be cost effective
- reduce working hours
- reduce stress and frustration
- reduce mistakes and problems
- eliminate significant waste from processes and daily operations (less *waste* = more *profit*)
- move from chaos to control
- continuously improve and stay relevant.

It is, however, easy to get carried away with standardisation and processes and I have seen this in some big organisations. Don't create processes for the sake of it. This does happen, unfortunately. Identify the jobs where you have clear inefficiencies or issues and start developing standard processes for these. Importantly, keep them simple and visual so that they are useful to everyone and easy to update and work with.

Lean Farm team quiz:

STANDARD WORK

1. What is standardisation?

2. What is standardisation the foundation of?

3. Who should be involved in standardisation?

4. What are the four questions you should ask before blaming a person for a mistake?

5. What are three benefits of standard work?

6. How can you sustain standard work?

7. What are three things you can standardise on a farm?

Lean Farm action plan:

Standard work

These actions are included to provide some simple guidance for your farm. They are aimed at giving you a little bit of motivation and direction if you need it. Of course, the idea is for you to do as much as you can to gain the maximum benefit of standard work on your farm.

1. Introduce standard work to your team.

2. Identify at least five tasks or processes on your farm that can be standardised.

3. Take *before* photos of any existing standards/processes.

4. With your team, video or document the current process.

5. Follow steps 1–11 in this chapter to develop your new standard process.

6. Train your team and roll out the new standard work document.

7. Take *after* photos.

8. Develop the necessary documentation, such as an audit sheet, to sustain the standard work.

Hopefully this information on standardised work has made you feel you can have better control over the outcomes of the jobs and processes that are done on your farm.

In chapter 8 we look at mapping your processes visually to help you see and implement any improvements needed.

Chapter 8

Value stream mapping

Value stream mapping (VSM) is based on a technique used at Toyota called Material and Information Flow Analysis (MIFA). It is essentially a process mapping technique with one key difference: it maps every step in the process—both value-adding steps and non–value adding steps (waste).

Most process mapping is in the form of diagrams with various boxes and shapes representing key process steps, decision points and flows. These mostly contain the value-added process steps and often don't give any information about waiting, walking, motion or other important data, such as the number of people doing a job and how long it takes. MIFA and VSMs in general, on the other hand, include this information.

This is why VSM is so valuable as a way of drawing a picture of any key end-to-end process. It is in this detail where the bulk of process improvement opportunities lie.

VSMs can be quite complex and technical, particularly when mapping complex manufacturing environments, for example, with shared resources, high variation, multiple product lines and product mixes.

I have adapted the VSM technique to make it far simpler for the purposes of dairy farming and this book. There is no need to complicate the mapping with a multitude of symbols and technical elements. The most important thing is that you actually put the process steps on a wall and start to discuss the waste so you can move forward and improve.

What is VSM?

A value stream is the complete collection of all activities or tasks in a process, from beginning to end, that are required to produce an outcome such as a product. It includes *all* activities, including those that add value and those that do not (waste).

The value stream map is a visual representation (diagram) of product flow through the selected process end to end at a given point in time. It captures and helps you to analyse activities, tasks and processes in your business visually, objectively and effectively. It is essentially a picture of your key business processes. Importantly, it is done *in the eyes of the customer*.

I have done hundreds of value stream maps with hundreds of businesses, including several farms, and it doesn't cease to amaze me what people who live in the process every day discover when they do this activity.

Value stream maps are used to identify potential problems and opportunities in an end-to-end process, whether it be a production, administrative or procedural process. Figure 8.1 demonstrates this (the 'W' in the triangle symbolises 'wait').

LEAVE PADDOCK   **RETURN TO PADDOCK**

Figure 8.1: an end-to-end farm process

Using the example in chapter 4, imagine if you put a red dot on one of your cows and you followed the cow through every single step, from the time it leaves a paddock until the time it returns to a paddock. You draw every one of these steps, including all the walking, waiting on platforms, waiting on yards, milking and eating on the feed pad from the start to the end and you time every single step. You also collect information such as the distance walked, the number of cows waiting at each point, the number of people working at each point, how information is flowing, vehicles used and so on. This is what a value stream map is. It is basically a drawing of all the elements in your process (imagine taking a birds-eye photo or video and then sketching it in 2D).

Value stream maps usually capture as a minimum:

- the process steps
- information flow
- the value (VA and NVA activities)
- process data
- process/people interactions
- any metrics or KPIs.

Key benefits of VSM

There are huge benefits to mapping your key farm processes this way. During my time with Toyota, as soon as I got onto a shop floor at one of the supplier factories, the first thing I would do is draw a MIFA. I used a pad of paper, clipboard and a pencil, we walked the process from start to finish (or backwards from customer end) and I drew everything I saw. At the end of the shop floor walk, I had a detailed and accurate picture on A3 paper of the entire factory operations, including all the problems I spotted. Having all the information on one page is incredibly valuable.

A visual process map will:

- help you see the waste in the process, quickly and easily
- help you capture and calculate the process lead time
- help you identify key improvements needed
- help you prioritise key improvement projects
- provide you with a snapshot of your process in the eyes of the customer
- identify your current operational thinking
- enable you to visualise how your product flows to the customer
- link your process information and material flows so you can view the process holistically
- establish a strategic improvement plan for your farm using a structured, fact-based, transparent process.

The process

There are five key stages to effective VSM (see figure 8.2). All five stages are essential if you want to improve your farm.

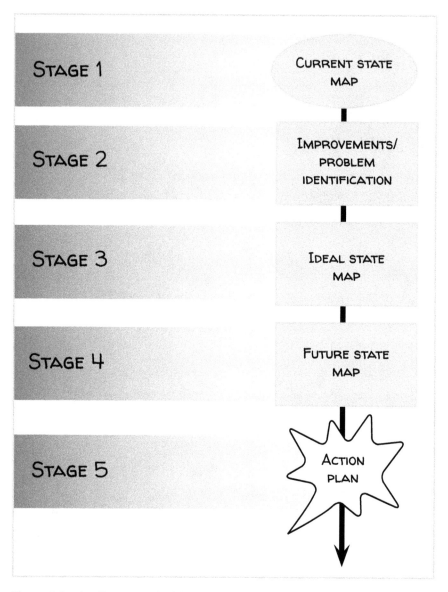

Figure 8.2: the 5 stages of VSM

Lean Farm team activity

Identifying a process

With your team, think about your farm processes.

1. What process could be a good process to map?

2. What would be the start and end point of this process?

Current state map

The first stage is the easiest—going out and mapping what is actually happening. Don't map what you think is happening or what is written on some old process document in the computer. Instead, walk the flow of the process from end to end, see it happening with your own eyes and record what is *actually* happening.

The correct way to map the process in a Lean world is to start with the customer first and map the process backwards. This, however, is often quite difficult for people who have never done it. I therefore recommend that you simply start at the beginning of the process, and just walk through as if you were doing it.

Going to see the process with your own eyes is called 'Genchi Genbutsu' in Japanese and is a critical element of a Lean culture. In English it is known as 'Go, Look, See'.

Improvements/problem identification

This phase involves looking at the visual map on a wall, discussing all the steps in the process and brainstorming all the problems and opportunities that exist within the process. It should be done with the whole team, so you can get full buy-in and everything is transparent. It is also the most enjoyable stage for the team where they can vent their frustrations.

Ideal state

Very often this stage gets left out when businesses do VSM. It is an important stage to think about. It involves mapping on a wall using sticky notes, or drawing or mapping what your selected end-to-end process would look like in the ideal world. This is the 'big picture' thinking that many farmers rarely find time to do.

The reason you'll want to map this ideal situation out, is that it helps you to create a vision of what your process would look like if there weren't any obstacles. While it may appear unrealistic today, and you may not even get there in the future, you can use your ideal state to drive the changes or improvements that you make today, so that you can at least start moving towards the ideal state. The ideal state is like your compass—it will guide you in the right direction, rather than making short-sighted or ad-hoc changes and improvements that won't take you into the future.

A Lean Farm example

Where did the time go?

A farm owner we did process mapping for had three dairy farms. The team from one dairy farm with a 520-cow herd did a process map of their afternoon milking process.

During the activity and discussion the team discovered that the process started at 3 pm and didn't finish until around 8.30 pm. This was approximately five and a half hours! The team compared this to another dairy farm with a similar shed and a herd size of around 600. That shed had a process time of approximately two hours.

By mapping out the process, the team could identify all the activities that they were doing, and in which sequence, and compare this to the other shed. This helped them to identify key improvement actions and work to reduce the process time to approximately three hours.

Saving: 2.5 hours of milking × 2 people = 5 labour hours per day

Future state

Unlike the ideal state, the future state is a map representation of what your process will actually look like once you have addressed some of the problems you have identified and made the necessary changes. It should be realistic and achievable. Once you implement your improvement ideas, you can re-map the new current state of the process and it should reflect this future state map. This is a good way to know that you have been successful with your actions.

Action plan

As always, there is no point doing any of these maps if you don't actually change anything in your process. The action plan is vital in helping you to agree on specific actions for each of the problems you want to address, and allocate these actions to people with a time frame.

Most farmers would agree that taking action is one of their biggest frustrations. Unfortunately, if you just expect people to remember actions or to do them with no follow up, it is likely that they won't get done any time soon.

The action plan formalises these actions and should be used regularly to follow up on the status of each action. This way people are accountable, and the actions can be discussed and are always front of mind.

Lean Farm team activity

The ideal state

Choose any farm process and with your team practise a blue-sky thinking activity. This basically means that you throw out all your preconceptions, think outside the box, are completely open to all possibilities and say that nothing is impossible. Once you are in this mindset, ask yourselves the following questions.

(continued)

Lean Farm team activity *(cont'd)*

— What makes this current process frustrating?

— What would this process look like if
everything was possible (e.g. no restrictions
in terms of cost, time or technology)?

Now draw your ideal state process.

VSM symbols

VSM uses many standardised symbols to represent elements in a process. They are well recognised among Lean practitioners and used in companies implementing Lean. You can do an entire week-long course on VSM if you want to know all the correct symbols and approach. I have only selected a few significant symbols, which are more than enough to get most farmers started in mapping their processes.

The important thing is that you use one symbol for waiting, one for process steps and one type of line to represent the product/process flow (see figure 8.3).

Current state map: the how

The first thing you will need to decide when mapping a current state is what process you want to map and what the start and end point of that process should be. You need to clearly define this start and finish point so that you are able to measure the process accurately each time.

The easiest way to think about doing a current state map correctly is, once again, to imagine you have put a red dot on a cow, a piece of paper or a person, and you follow that red dot through every single step from the start to the finish. Remember: you are drawing the *actual process*, not what you think the process is or what is drawn on some existing document. Believe me, this will be very different from what you actually see.

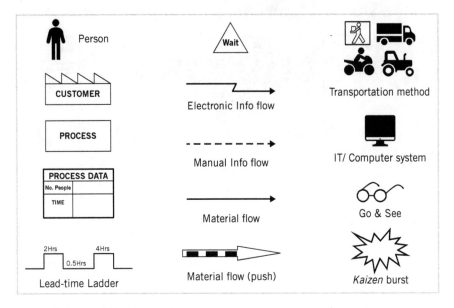

Figure 8.3: key VSM symbols

Don't worry if you can't complete all of these steps. It doesn't need to be perfect. The most important thing is to visually show your key process steps with an estimated time. If you can get to this point, it will already be highly valuable.

Step 1: walk and understand the actual process

I have provided some steps to help you do this as effectively as possible.

1. **Walk the process from start to finish.** Do this with the team if possible so that everyone can get a feeling for what is actually happening.

2. **Sketch every key step, every wait, everything you see happening on paper—use a pencil.**

Cows waiting on the feed pad before milking

3. **Talk to people doing the job if you need to, without distracting them too much.** Ask them questions such as:

   a   How long does this normally take?

   b   How do you know what you need to do next?

   c   How often does it need to be done?

   d   When does it need to be done?

   e   Where does it need to be done?

   f   Who does it?

   g   Why is it done?

   h   How does it need to be done?

   i   What information do you need?

   j   Where does the information come from?

4. **Capture data where possible. The type of data you can capture includes:**

   a   how long each step takes

   b   how long a cow or person or product is waiting for something

   c   the number of cows in one place or in a queue

d   whether there are defects or quality issues in the process

e   the number of people needed for a particular task

f   the method of moving (e.g. tractor or bike)

g   how much you need to produce and when.

5.   **Identify any problems or 'kaizen bursts'**

a   Take note of any problems you see as you are walking and observing the process.

b   Draw a *kaizen* burst symbol on your hand-drawn process map/ sketch indicating where in the process the problem is and make a note of what the problem is.

Step 2: create the current state map

Once you have completed your sketch of the process and you have captured as much data as possible, you should get the team together and start to put the current state map together on a wall.

What you will need:

• butchers paper — ideally a roll

• sticky tape/painters tape (or similar) to prevent wall damage

Cows waiting in the yard before being milked
Source: Grant Matthew

- sticky notes in multiple colours

- a pencil, ruler and eraser

- your process sketch

- your process data

1. **Draw the process steps**

    a   Stick a large sheet of butchers paper on a wall.

    b   Use a different colour sticky note for different people/roles working in the process and for 'waits' (this helps to visually identify which roles do what in the process).

    c   Develop a sticky note colour 'key' so that everyone remembers which colour represents what.

A tip: I try to always use bright pink for 'problems' or '*kaizen* bursts' identified in the problem brainstorming rather than for process steps — this way they stand out. I also always use either bright yellow or orange for the 'wait' steps as these colours also stand out.

Mapping the milking process

d   Write each process step on one sticky note.

e   Place the sticky notes in sequence on the butchers paper.

f   Try to keep the sticky notes aligned to other simultaneous steps and in a time-based sequence.

g   If there were any waits in the process, draw a 'wait' symbol (triangle with 'W' in it) and write below it what/where the wait is.

h   Add the 'wait' sticky notes on your map at the point where they occur in the process.

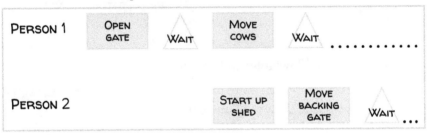

2. Add process data

Under each key process step, add any data you have managed to obtain related to that step. This could be the number of people working on that step, frequency, quality percentage, cow numbers and so on.

3. Add times for every process step, including the 'waits'

Under each sticky note, record the time it takes to do that step in the process. It's good to use the same time unit (e.g. minutes). Make sure you also record the time for each wait—how long that wait was.

For example, if a person was standing and waiting for five minutes before they could do their next process step, write '5 min' below the 'wait' sticky note.

PERSON 1	OPEN GATE	WAIT	MOVE COWS	WAIT	. . . .
	3 MIN	15 MIN	36 MIN	47 MIN	

PERSON 2			START UP SHED	MOVE BACKING GATE	WAIT	. . .
			24 MIN	4 MIN	2 MIN	

4. **Add 'material flow' lines to link each process step**

Join each of the process steps and waits with an arrow to visually show the flow of the process. This just makes your process clearer to everyone. If there are steps that repeat or flow in two possible directions, add the arrow to show a loop around or the new direction.

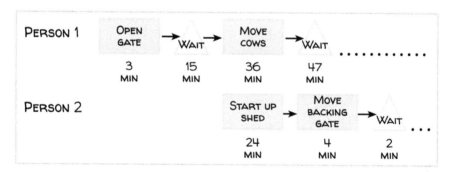

A Lean Farm example

It's all about the time

Another farm had visualised the simultaneous tasks that several people were doing as part of the milking process against a timeline. This visual map showed that two people were arriving at work and starting their plant start-up too early, resulting in them having about a 30-minute wait until the cows were in the yards ready to be milked. Not only was this a waste but it also meant that their work day was unnecessarily long.

Saving: 30 min × 2 people per milking = 2 labour hours/day

The team agreed that those doing the plant start-up process would start work 30 minutes after the person bringing the cows in. This eliminated the unnecessary waiting and early starts.

5. **Add information flow**

   Electronic information flow: If you use a variety of software, such as herd management, pasture management or production software, for example, this will usually create some electronic information flow. You can draw a computer on a sticky note and name it as per the software system. You then draw the zigzag line between the computer sticky note and the process step which needs to either input information into the system (arrow points towards computer) or extract information out of the system to do a particular task (arrow points towards the process).

   Manual information flow: This is information that is manually conveyed either by telephone, speaking physically with another person (e.g. meeting or verbal instructions) or using a book/procedure manual or similar. This type of information flow is very common on farms and is illustrated as a dotted line flowing between the sources of information and the process step.

   The main reason you want to draw your information flow is that it tells you how people get information, and from where, to be able to do their jobs. If you start to get a lot of information flow lines darting back and forth all over your map it gives a strong indication that your process may be very inefficient, complicated and chaotic. Communication is vital to have a productive business, and if there isn't clear, smooth flow of information, your process is probably not as productive as it could be. You may also be seeing many mistakes or problems, which could be a result of poor information flow. If your information flow looks complicated, it probably is, which may affect communication.

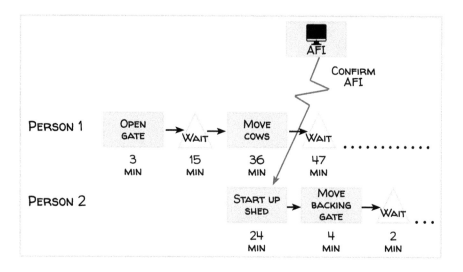

Lean Farm team activity

Genchi Genbutsu

Choose a process to follow with the team.

1. 'Go, Look, See' with the team—follow the farm process physically from start to finish.

2. Sketch the process and time each step.

3. How much waiting is included in this process?

4. Ask each team member to guess the value-added percentage in the process.

5. Calculate the value-added percentage and discuss the result with the team.

6. **Draw a 'lead-time ladder'**

This is a key feature of the value stream map. The lead-time ladder tells you what the total end-to-end process lead time is in the eyes of a customer. It also tells you how much of the end-to-end process lead time is waste vs value added. It looks like a step ladder: the step goes down if there is a process step and up if there is a 'wait' sticky note. We know that 'wait' is a waste, which means all the 'up' steps in the lead time are automatically waste. You can then very quickly calculate out of the total lead time, how much time is 'wait' time.

Mat rounding up the cows for milking

A current state process map of the milking process

The down steps on the lead-time ladder are processes or actual tasks and these processes will be either value-adding or non–value adding.

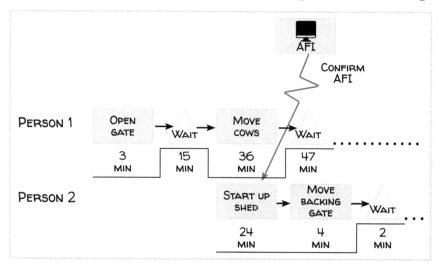

7. Decide which 'process steps' are value-adding

a   With the team, question your process steps and ask if a customer would pay for you to do that particular process step. If yes, then it is a 'value-add' step. If no, then it is a non–value add step or, in other words, waste.

b   Mark all the 'waste' steps with a red dot.

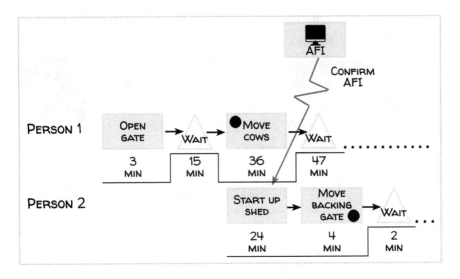

8. **Calculate the total process lead time and the total value-add in the process**

    a   Add up the total lead time (the times below all the sticky notes) and write this total at the end of the process map in a box.

    b   Add up all the times for the value-added process steps only.

    c   Divide the value-added time by the total lead time — this will give you a percentage value add of your end-to-end process in the eyes of the customer.

    Note: if you have multiple branches of process steps because of simultaneous activities, for the purpose of this activity identify what the critical path is and use that as your total lead time.

    Note: the way I have asked you to calculate the lead time and process time above is not the way it is done in the true Lean world. I have simplified it so that it is easy to do and relevant to farmers rather than a bunch of Lean terminology!

$$\frac{\text{VA PROCESS TIME}}{\text{TOTAL LEAD-TIME}} \times 100 = \%VA$$

9. **Put your current state map into an electronic format**

Finally, you should keep a good electronic record of your process map in case the paper version gets destroyed or lost. You can do this by simply taking a photo, or using Excel or a process mapping software package such as Microsoft Visio or SmartDraw.

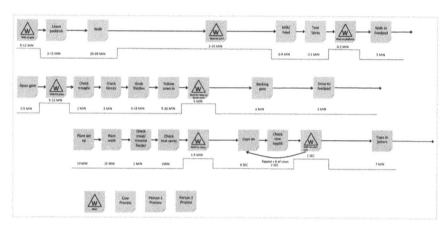

Example of a visual map

Improvements/problem identification: the how

This stage of VSM is the most exciting and the whole team should be involved.

Step 1: open brainstorm (your kaizen bursts)

To initiate this activity:

- Ask the team to brainstorm any problems or frustrations they currently experience with the process. The team can think about and record the problems individually or discuss them with others.

- Ask each person to record their problems on pink sticky notes (this is the *kaizen* burst) — one problem per sticky note in legible writing. (Try to make the problem as specific as possible so people know what it means later when you are reading through the problems.)

- If the problem is related to a specific step in the process, add the pink sticky notes next to that part of the process. If it is a general problem (e.g. communication, IT system) place the sticky note somewhere in a free spot on your map.

A Lean Farm example

Defeating the 2 am start

The process map can help to visualise very quickly a variety of problems. One farm manager mapped their morning milking process, which started exceptionally early (at 2–2.30 am). This meant extremely long work hours for the manager, which no-one was aware of. The process map helped to drive the right analysis and discussion to understand what was going on and work to push the starting time back to a more acceptable time.

Saving: potential reduction of work day by 1.5–2 hours

Step 2: the 8 wastes

To identify the 8 wastes:

- Ask the team to think about the 8 wastes (which you learned about in chapter 4).

- Identify any of these 8 wastes in your process (you should already have all the 'waits' in bright orange or yellow and know these are wastes). This could be extra walking, moving, over-processing, rework and so on. Write the waste on a pink sticky note and stick the sticky note on the map next to where you found the waste.

Cows walking to a dairy shed

Step 3: other opportunities

Consider other possibilities.

- Ask the team if they have any other ideas about where the process can improve that haven't been covered in steps 1 and 2.
- Record these on the pink sticky notes and add to the process map.
- Once you have completed all the brainstorming, and all your sticky notes are on the wall, take a photo.

Step 4: organise

To organise:

- Number all the problems from 1–n.
- Run through all the problems identified with your team and try to group any that are similar or duplicates together into one group.

- Decide which problems are feasible to action and can be tackled either immediately or in the near future. Focus on these problems first.
- Start to think about what actions could be taken to address the problem identified. Record these actions on a flipchart with a reference to the original problem number for easy tracking.

Discussing the current state process

Here are some questions to ask when you're finding opportunities in the current state process:

- What are the customer requirements?
- How can we make this simpler?
- How many people are involved?
- What does the data tell us?
- How many repeated or duplicated steps are there?
- Why are there delays?
- What is the flow like? Is it streamlined?
- Where are the bottlenecks in the process?
- What process steps do not add value?
- What is the overall lead time?
- Where are the problems?
- Where/What are the wastes?
- What improvements can be made?
- Is there potentially a better layout/flow?

Ideal state: the how

The ideal state should be a process that is highly streamlined, very quick and efficient, needs almost no prompting, and delivers a reliable, perfect product as needed every time at the lowest possible cost. It should only consist of value-added process steps—ones the customer would pay for. It should be faultless: there should be no mistakes or problems and it should be highly adaptable. Sounds like a dream, right? Now try to design it.

REMEMBER ABSOLUTELY EVERYTHING IS POSSIBLE.

- Set up another piece of butchers paper on the wall. Ask the team to brainstorm what an 'ideal' process would look like. This may mean using robotics, special IT, drones, changing layouts, new sheds and so forth. Encourage 'out of the box', completely open thinking. Even if it sounds crazy, discuss it and consider it.

- When preparing this ideal state, think about how long the current process is and all the waiting and non–value added steps. What would the ideal state look like if you simply got rid of all the waits and non–value add? How could you do this?

- Map the ideal state using sticky notes.

- Add a potential lead time ladder if you can.

- Put your ideal state map into an electronic format by taking a photo, or using Excel or a process mapping software package such as Visio or SmartDraw.

Now you have a vision. This ideal state may not be achievable right now or in the near future, but if this is what your ideal state would look like, you know what kinds of things you should start trying to improve and change.

Future state: the how

Now that you have a vision, you can get back to reality and design a realistic future state that is aligned with your vision or ideal state and moves you at least a little towards this ideal. It could be eliminating waits, changing a layout or simply removing steps that are just unnecessary. This is all reducing our process time and helping us move towards the ideal state.

The real focus of the future state should be trying to eliminate as many 'waits' as possible and making the process more streamlined.

- Set up another butchers paper on a wall.

- Review your problems and the improvement actions you have identified to solve the problem.

- With the team, discuss how the process would change if a specific improvement action is implemented.

- Use sticky notes again to draw up your realistic future state map based on the immediate or shorter/medium term improvements you will make. Use one sticky note again for each process step. It will be likely that some of the 'waits' you have in your current process will remain as you can't eliminate them in the near future. If this is the case, include these in the future state map also.

- Put your future state map into an electronic format either by taking a photo, using Excel or a process mapping software package such as Visio or Smartdraw.

Action plan: the how

The team has done a lot of work to establish a better future state for the farm. All this hard work can go to waste if there isn't a clear action plan that is documented and everyone is committed to.

Step 1: agree on an action plan

- Draw up an action plan template on a flipchart or whiteboard (see figure 8.4, overleaf). This should include a column for what needs to be done, by whom and by when.

- Finalise all the actions you want to implement to achieve your future state.

- Record each of the actions you have agreed to on your action plan and link it with the problem number from the pink sticky note that this action will solve.

- Allocate each action to a specific team member to lead that action and take ownership of its implementation — let the team decide which actions they will own and allocate their name against the action.

- Agree on a time frame for each action with a specific date by which the action should be closed. Add this date to the action plan.

ACTION PLAN				
#	WHAT	WHO	WHEN	STATUS
1	AUTOMATIC GATE OPENER—TRIAL FOR AM MILKING	JS	10/06	◐
2	FIX RACES BY PADDOCKS 40–44	RK	20/07	◔
3	BACKING GATE OPERATION FROM CUPS ON AREA—ADD CONTROL	MH	30/06	◔
4	FEED HOPPER AT BODMIN NEEDS REPLACEMENT	JS	20/07	◑
5	CLEARER PROCESS REQUIRED FOR FIXING TROUGHS	SC	19/06	◕
..	....			

Figure 8.4: an example of an action plan

Step 2: follow up

Once your action plan has been documented and agreed on you will need to regularly follow up on the actions to ensure there is progress.

Make this an agenda item at your weekly team meetings, so that you can build a disciplined follow-up process. If actions are not on track, the team needs to explain why and decide on a countermeasure to catch up.

Setting up measures that can help you to see if the action has been effective is a good way to validate your improvements and showcase tangible results.

Step 3: celebrate!

It is always important to celebrate your success. Once a few of the quick wins have been implemented ensure that the team is acknowledged for their hard work. Publicise the achievement and improvement so that everyone can see the difference their effort has made to the business. Reinforce the team's important contribution and do something small as a token of appreciation.

Continuous improvement

Once you have actioned all the items on your action plan, your new process should look like the future state map that you drew originally. A good way to test this is to walk the process again and map it out once more. This is your new current state map. If this new current state map looks like the future state that you had drawn, then well done! You have made a big improvement in your process. This is a great visual way for the team to see the difference their actions have made to the original process.

The improvement shouldn't stop though! You now have a new current state map. By repeating this entire VSM process every year, you can continue to refine and improve the process until you eventually achieve your ideal state. This is the cycle of continuous improvement discussed in chapter 2.

Summary

VSM is a highly visual, interactive tool that will help your farm make real, tangible improvements in your processes. It is a key Lean tool used by all sorts of businesses to drive strategic, structured and incremental improvements.

The value stream map visualises a core business process in the eyes of the customer. Importantly, it shows the *actual* state: what is really happening in the process, rather than what people think is happening.

VSM helps your farm to identify and differentiate between value-adding and non–value adding (waste) steps in your process and to see where the waits are.

Lean Farm team quiz:

VALUE STREAM MAPPING

1. What does VSM stand for?

2. What is meant by current state VSM?

3. What are the five key steps to VSM?

4. What four things does a VSM show?

5. Draw four symbols used in VSM.

6. What is the lead-time ladder?

7. How do you calculate the value add in the process?

Lean Farm action plan:

VALUE STREAM MAPPING

These actions are included to provide some simple guidance for your farm. They are aimed at giving you a little bit of motivation and direction if you need it. Of course, the idea is for you to do as much as you can to gain the maximum benefit of VSM on your farm.

1. Introduce VSM to your team.

2. Identify at least two key processes on your farm.

3. Use the correct steps to 'walk' the process.

4. Create a map of the actual current state.

5. Take photos of your current state map.

6. With your team, brainstorm current problems, waste and opportunities in your current state process.

7. Take photos of your problems on your map.

8. Organise your problems and identify specific actions/ improvements.

9. Design and map your ideal state process. Take photos.

10. Design and map your new future state process. Take photos.

11. Develop a clear action plan to implement necessary improvements that will achieve the future state.

12. Implement any 'quick wins'.

13. Follow up your action plan regularly and celebrate success.

VSM will help you look at your process in a different way and identify many problems. Some of these problems may not be easy to solve and you will need to dig further to truly understand them. The next tool I am going to introduce is practical problem solving. This tool will help you better understand and solve any problems.

Chapter 9

Practical problem solving

As farmers, we are pretty good at problem solving, right? Particularly with the 'number 8 wire' approach. But do we see the same problems popping up again and again? The number 8 wire approach is fantastic if we have an urgent situation and need to put in a fix right then and there. But if we spend our life going around number 8 wiring, then we end up with a lot of bandaids all over the farm that we keep having to go back to repair or fix over and over again. Furthermore, if you are seeing the same type of problems reappearing, it tells me that you haven't fixed the root cause of the problem yet in order to prevent it from happening again.

In this chapter I am going to give you and your team a couple of tools to help you problem solve in a more structured way so you can find the correct root cause of a problem and fix it for good rather than just bandaiding all the time.

We *love* problems

Before we get started, have you ever heard of this saying or used it yourself?

No news = good news

From today on, I want you to turn this around to:

No news = bad news

This is because there are always problems out there—you just haven't heard or found out about them yet. But they are there, waiting quietly until they get big enough to explode. Therefore, the more we can encourage our team and ourselves to identify, raise and talk about problems, the more likely we'll know about them, and solve them quickly before it's too late. Furthermore, if you keep telling your team or colleagues 'no news = good news', you are actually instilling the mindset that you don't want to hear about their problems and that problems are a bad thing. This will make your team too scared and uncomfortable to raise any problems or anything that goes wrong as they will not want to upset the boss. This means that if something does go wrong no-one will own up to it or raise it out of fear that the boss won't be happy. Obviously, this type of behaviour is not very conducive to having a well-functioning, preventative and efficient farm. In the words of Taiichi Ohno, **'No one has more trouble than the person who claims to have no trouble.'**

For you to be able to get problems out in the open, talk about them and proactively fix them, you and your team need to have a positive attitude towards problems. Problems help us to find and eliminate waste from our farm and we should not be afraid of them or avoid them.

Not raising problems early enough can lead to problems such as incorrect grazing and feed
Source: Grant Matthew

When we know about problems, we can do something about them and continue to improve our business. When we don't know about problems until it's too late, we cannot improve, but instead spend our days firefighting. We therefore need to change our attitude towards problems so they are seen as a positive rather than a negative thing. Again, if your farm has a negative attitude towards problems, and a Rottweiler hounds someone down if they raise a problem, then the chances of that person raising a problem in the future are close to zero. While it might be tempting to get very worked up over an issue, it won't help your business in the long run. All you will do is create fear among your staff and no-one will ever raise any issue as they will be too worried to upset the boss, and it will be more comfortable for them to remain silent.

Have you ever heard of the following saying or said it yourself?

'Don't come to me with the problem, come to me with the solution.'

This is the second phrase I want you to delete from your mind and never use again. Instead, from today, whenever someone comes to you with a problem or a solution you are going to ask them:

'WHAT IS THE PROBLEM?' AND AGAIN 'WHAT IS THE PROBLEM?'

What you will realise is that if people have just jumped to a solution without correctly understanding what the problem is and what is the root cause of the problem, it is likely that the solution they have put in place won't actually solve the problem for good. Therefore, before putting in place any solution you need to make sure that a) you truly understand what the problem is and b) you have got to the *real* root cause of the problem. Furthermore, sometimes what people think is a problem might not actually be a problem. By asking 'What is the problem?' you might establish that there isn't really a problem to start with, or what people think is the problem isn't the real problem.

Lean Farm team activity

Problem solving on the farm

With your team, discuss the following:

1. How do you currently solve problems on the farm?

2. Who is involved?

3. Where do you do it?

4. When do you do it?

5. How do you do it?

When a problem becomes a *problem*

Most problems start as a small, potentially trivial, problem somewhere. If you don't come across that problem as soon as it happens, or no-one tells you about it, as time goes on the problem grows bigger and bigger. By the time you do finally discover that there is a problem it has often turned into a *big* ugly problem that is a lot harder and more costly to fix.

Figure 9.1 is a good way to think about problems. The quicker we can identify them, the cheaper and more easily we can fix them.

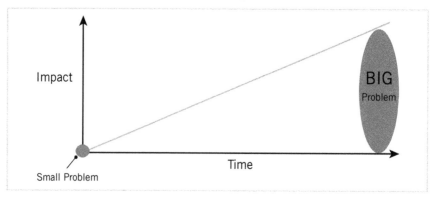

Figure 9.1: problems grow bigger over time

The key thing is that you really need to create an environment on your farm where people are proactive in raising problems and feel comfortable doing so. This way, if they see a problem or something goes wrong they will immediately raise it rather than trying to hide it or ignoring it.

Lean Farm examples

Don't let small problems become big ones!

How many times have you had a motorbike that was low on oil? It starts as a small problem. But what happens when no-one says anything or checks it? Eventually the bike has no oil and it seizes. You have potentially caused some big damage to your bike. The cost of repairing a motorbike or, worst-case scenario, having to buy a new one is a lot greater than buying oil, isn't it?

Brainstorming problems identified during calving season

I am sure we can all think of multiple examples of problems that could have been fixed much sooner with less effort if someone had known about them. So why didn't anyone raise the problem? It is a good question to ask yourself and your team to help you be more proactive about problems. Here is another example from our farm.

Lean Farm examples

Musical paddocks

This is a situation that took place on our farm.

Person 1 was responsible for putting cows into a new paddock. He put them into Paddock 1 and left, but forgot to check the gates were all closed. One gate at the back of the paddock was left open. Of course, the cows saw that the grass was greener on the other side and all raced into Paddock 2 next door. Person 2 came driving by some time later and saw what happened so he moved them back into Paddock 1 and let Person 1 know. Of course, as the cows had had a good munch on Paddock 2, they had left Paddock 1 poorly grazed. The manager had a day off but happened to drive past the cows and noticed how poorly grazed Paddock 1 was. As they were post calving, his priority was to ensure the cows were well fed. Seeing the poor grazing he thought that perhaps something was wrong with the grass in Paddock 1 that was stopping them from eating it.

So he devised a whole new grazing plan and got Person 1 back to move the cows to a new paddock—Paddock 3. So now three paddocks had been affected. Person 1 still didn't say anything about the original problem: that the gates had been left open and the cows had gone into Paddock 2. Eventually, the next day, the manager found out what had really happened. Obviously, if the problem had been raised straight away, it would have saved one paddock from being grazed unnecessarily, and a lot of additional time and resources developing new grazing plans and moving cows around.

What is a problem?

A problem is essentially a gap between the current situation and your target or ideal situation.

There are two main types of problems that we deal with on farms: something I call Level 1 type problems and Level 2 type problems.

Level 1 problems are those that we deal with most often—our current performance is below our set standard or expectation.

Level 2 problems are those where we are already performing to our standard or expectation but we want to achieve even more and have decided to set ourselves a higher ideal target.

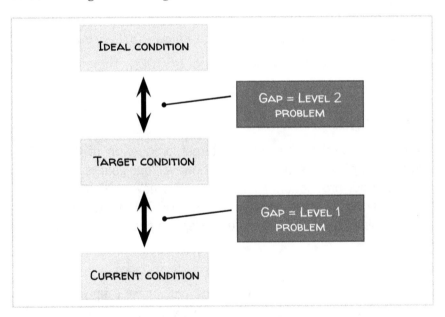

The *same* problem again?

If we started collecting information on every problem we have on our farm, how many problems do you think would be similar to ones you've had before, or exactly the same as problems you've had before?

From my discussions with many farmers, all have agreed that in fact most problems they see fall into one of the two groups: either they are similar to or exactly the same as a past problem. And if we really did capture data about our problems, we would discover that this is in fact the case. Many of our problems are repeat problems (see figure 9.2).

Lean Farm examples

Level 1 type problems

— Our standard is that everyone closes farm gates, but we are constantly dealing with gates being left open. There is obviously a gap between our standard and our actual situation.

— Our target is to have a cell count below 200000 but our current situation is a cell count of 350000. Therefore, our problem is that we have a gap of 150000.

Level 2 type problems

We are currently achieving our target production of 400000 kilograms of milk solids (KgMS); however our goal is to try to improve our production, so we have set a new stretch target of 500000 KgMS. Our problem gap is 100000 KgMS. Level 2 type problems are usually more strategic types of problems.

Why do you think this is? To me it appears that either we don't have time to fix the problem properly the first time or we simply haven't got to the true root cause of the problem so that we can put in place a proper countermeasure that will eliminate the problem for good. Unfortunately, if we are spending all day fixing problems that really shouldn't have occurred in the first place, or that will end up re-occurring in the future, then we are really wasting a lot of time and money.

So how do we currently fix problems on the farm? From what I have seen on our farm we spend a lot of time running around like headless chickens from one problem to another, putting out fires and slapping on bandaids (or no 8 wiring) and then having to quickly move on to the next thing. Ideally—and this is what this chapter is all about—we need to become proactive about problems and use a better, systematic approach to solving them so that we can start to get rid of repeat problems for good. See figure 9.3 (overleaf) for a comparison of traditional and Lean approaches to problem solving.

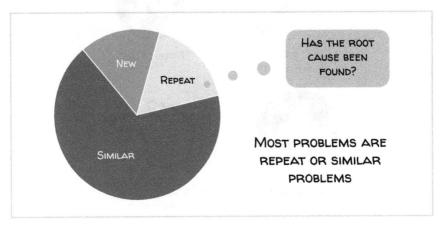

Figure 9.2: types of problems

Lean Farm examples

Power on!

During a problem-solving workshop, a farmer was telling me a story about how their fence electricity power flicked off whenever there was a power cut and it didn't re-start again when the power was back. Most of the time no-one knew that there had been a power cut, so the fence stayed without electricity until cows broke out and someone realised that there was no power and switched it back on. This had been happening for six years! Imagine all the lost time and problems!

The team had never sat down and used a structured problem-solving tool to try to get to the bottom of the problem. Instead, they kept bandaiding and having the problem re-occur. That was until they did my LeanFarm workshop, of course!

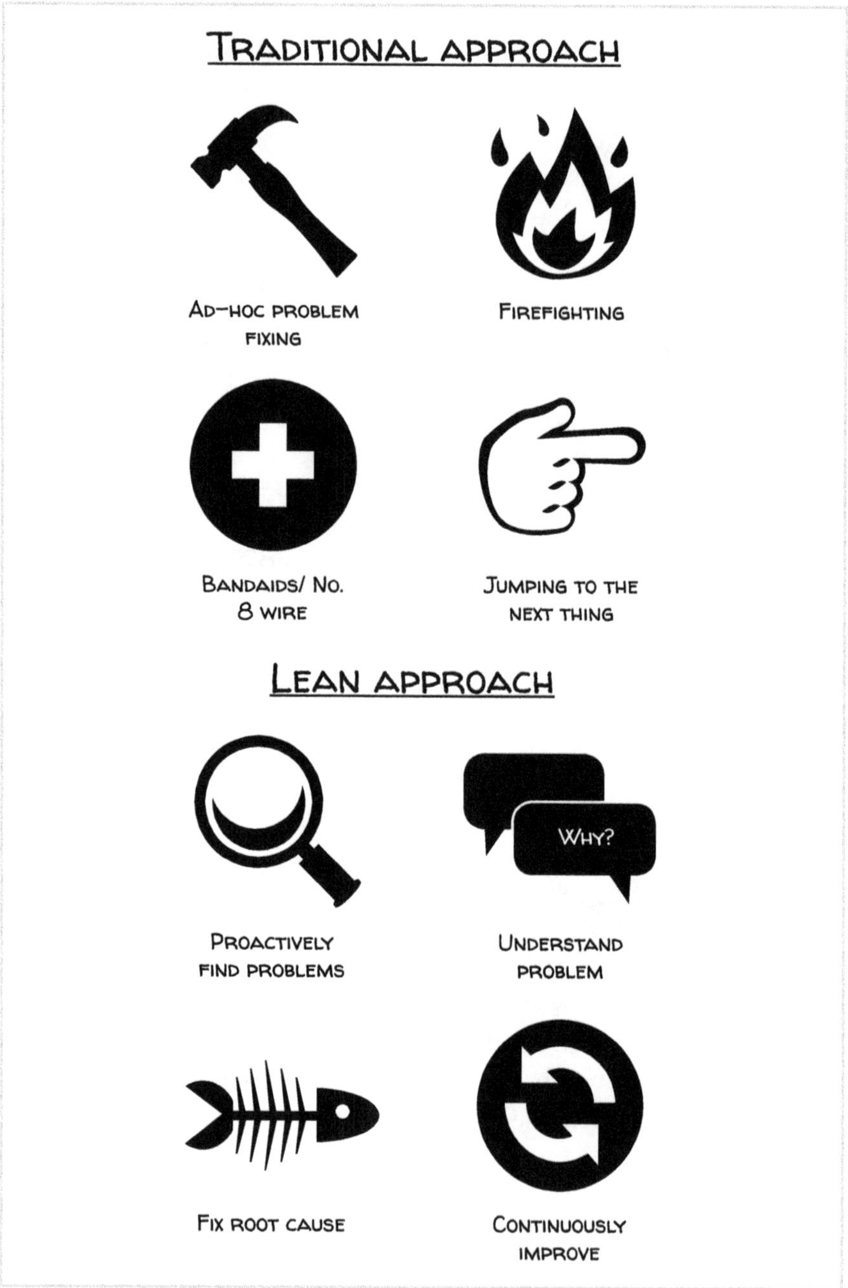

Figure 9.3: our approaches to problem solving

Lean Farm team activity

Not again!—repeat problems

With your team, brainstorm problems that you have recently had on your farm.

1. Are there any problems that you have already seen before? List them on a wall chart.

2. Why do you think they have happened again?

3. How were these problems tackled in the past—traditionally or in a 'Lean' way?

4. What did you do to try and fix them?

What makes a good problem solver?

Most of us probably think we are good problem solvers. Unfortunately, as we have discussed, there is a big difference between truly solving a problem to the root cause so that it never occurs again and inventing a fabulous bandaid to hold the problem together until you have time to fix it properly (which often doesn't eventuate, let's be honest).

Here are a few things that make better, more strategic and structured problem solvers:

- Constantly question the purpose: *why?*
- Start by understanding the *real* problem.
- Look at things from a customer perspective.
- Always obtain facts and make *fact*-based decisions.
- Never jump straight to conclusions or solutions.
- Take ownership of the problem-solving—don't expect an answer to be handed to you.

- Make information—and the process—visual.

- Confront problems with an open mind and ask 'How can we do this?'

- Be thorough in following actions through.

- Be prompt and proactive.

- Involve the team to achieve better outcomes—you are more likely to come up with the right solution with the team than on your own.

Figure 9.4 sums these attributes up succinctly.

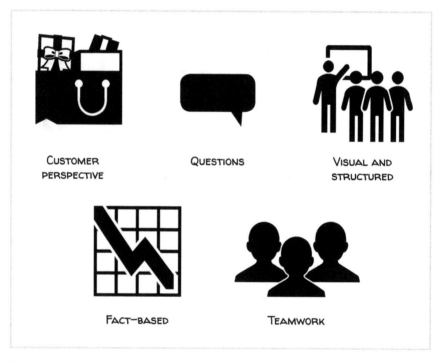

Figure 9.4: attributes of a good problem solver

Problem-solving tools

At Toyota, the Toyota Production System (TPS) is based around having people who are strong problem solvers. Everyone in the business is taught problem-solving tools and techniques, and involved in problem-solving processes. Promotions (at least when I was at Toyota) are based on demonstrating your ability to solve complex problems and you must complete a practical problem-solving project to prove your ability.

Having a few simple problem-solving tools and systems in your toolbox can help your team take a more structured and effective approach to problems. A structured process to problem solving is very important as it will move your farm from the 'traditional approach' to the 'Lean approach'. Without a structured process it is easy to:

- ignore a problem or not see it at all
- solve the wrong problem as you didn't understand it in the first place
- solve only part of a problem
- solve a problem that isn't actually a problem
- solve one problem and create a different problem
- put the problem into the too-hard basket and just put up with it
- not solve anything and put in place a bandaid instead.

There are four key Lean problem-solving tools that are commonly used:

1. the 'problem and countermeasure' sheet
2. the 5 Whys
3. the fishbone diagram (also known as the *Ishikawa* or 'cause-and-effect diagram')
4. the Lean structured practical problem-solving method (used by Toyota and other Lean companies).

These are illustrated in figure 9.5 (overleaf).

In this chapter I will explain the first three tools. These are very simple and anyone on the farm can use them to help solve most problems at any time.

In the LeanFarm workshops that I run with farms, I include a module on the structured practical problem-solving method. However, I have found it is too clunky for the majority of farmers, with most finding it difficult to follow and overwhelming. I have therefore decided to exclude it from the book in order to keep the content as simple as possible.

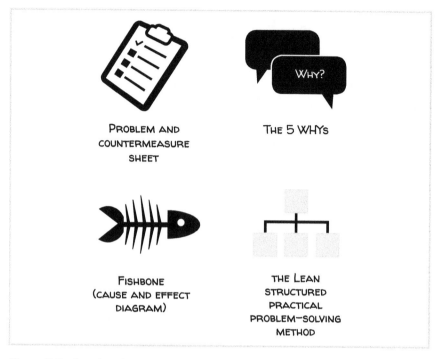

Figure 9.5: four key Lean problem-solving tools

Examples of common problems on the farm where you can use problem-solving tools such as the 5 Whys and the fishbone diagram include:

- cell count too high
- coliform or other grades
- cow mix-ups
- no water in troughs
- no fence electricity
- production down
- high level of mastitis
- plant not identifying cow number correctly
- feed hopper not allocating correct feed amount
- cows being missed on platform
- computer not capturing data
- cows not feeding well
- high maintenance costs
- broken equipment
- broken motorbikes/quads
- water tanks leaking
- no hot water
- high lameness in cows
- large quantity of downer cows

- high empty rate
- high vet costs
- turnover of team members
- too much bought-in feed.

'Problem and countermeasure' sheet

The first thing every farm needs to introduce is some sort of tracking of problems. If problems are not recorded somewhere, it is very easy to forget about them — particularly if your memory is anything like mine. Either you forget whether you have actually done anything to fix them, the person who was supposed to fix it doesn't remember that it was their responsibility, or the problem turns big and ugly and comes back to bite you in the backside when you least expect it.

Having a standard problem and countermeasure sheet in place helps you to record your problems immediately (as soon as someone raises something). You can then identify a suitable solution (in engineering terms, a countermeasure) — which could be once you have done some root cause analysis — and allocate a person and time frame to the action.

The problem record sheet helps you remember all of your problems so that none can be forgotten, and it is a good document to use to track the status of the countermeasure on a regular basis (such as during your team meetings).One of ours is shown in figure 9.6 (overleaf). Other benefits include:

- it helps you to work through the problems one by one, fix them for good and improve your farm
- by solving problems for good, your day will be more productive (instead of being spent running around putting out fires)
- the sheet helps to visualise progress so the team can see their achievements
- it ensures problems are owned and solved by the whole team, not just the manager
- it provides transparency of all issues
- problems aren't hidden, ignored or forgotten.

Figure 9.6: filling out our 'problem and countermeasure' sheet

Figure 9.7 is a worked example of a problem and countermeasure sheet that you can use to start to record your on-farm problems. You can also get this template electronically from www.leanfarm.nz.

The 5 Whys

The 5 Whys is an ingenious tool—simple and effective. Everyone can do it. In fact, you were already doing this when you were a child. If you have ever been around small children you might know that they ask 'why?' about everything. You respond and then they ask again, 'but why?' and you need to find another explanation. This is actually a very valuable technique that is used by some of the best companies globally to get to the root cause of a problem.

REMEMBER WHEN YOU WERE A CHILD
AND YOU ASKED 'WHY?' ...

PROBLEM REGISTER

#	TYPE	PROBLEM/CONCERN	COUNTERMEASURE	WHO	WHEN	RESULT	STATUS
1	Process	Only 1 person knows how to set up irrigator	Do e-learning for irrigator set up so anyone can do	HB	11/04	E-learning done and saved on sharepoint	 ◕
2	Facility	Mineral pallet is too far away	Set up mineral shed next to each silage stack	AR	30/05		 ◒
3	Facility	Water leak under platform	Loop piping overhead and block original pipes	MH	18/04	Pipes brought overhead	 ◕
4	Farm	Power loss in paddocks 30–34	Investigate where power is lost and fix	JW	30/04		 ◒
5	Facility	Cows not getting off platform	Do brainstorm around different options to attract cows off platform	JS	30/04		 ◔
...							

Figure 9.7: a worked example

The concept behind the 5 Whys is that if you have a problem, rather than jumping to the solution (which we are all guilty of), you should stop and ask yourself 'why' this happened and ask this five times. The number five has been used as this is generally how many times you need to ask 'why?' to ensure you get to a root cause. However, sometimes you may only need to ask 'why' twice and other times you may need to do it seven times.

Lean Farm examples

Getting to the root of the problem

Here is an example where the plant stopped working and no-one knew what had happened. A huge amount of time was spent trialling all sorts of things to get it running again — none of which solved the root cause. Eventually the cause was discovered by accident. If the 5 Whys had been used, the team could have identified the cause of the problem more quickly and not wasted significant time and energy.

Problem to solve: The plant stopped operating

1 WHY?	Why did it stop?	The fuse melted _THEREFORE_
2 WHY?	Why did the fuse melt?	Circuit overloaded _THEREFORE_
3 WHY?	Why did it overload?	Bearings were damaged and locked up _THEREFORE_
4 WHY?	Why were the bearings damaged?	Insufficient lubrication _THEREFORE_
5 WHY?	Why was there insufficient lubrication?	The oil pump not circulating oil _THEREFORE_
6 WHY?	Why was the oil pump not circulating oil?	Pump intake was blocked _THEREFORE_
7 WHY?	Why was the pump intake blocked?	There was no filter on pump intake

As you can see, this simple yet powerful 5 Why tool can help your team think more deeply about a problem and solve it more effectively and quickly. This will save your team time and your business money.

Lean Farm team activity

The 5 Whys

1. Think of a problem you had over the past week.

2. Use the 5 Whys to find the root cause of this problem.

Richard doing a 5 Why on why cows weren't being fed on the platform

Fishbone diagrams

While this tool is part of the Lean tools, it actually is a tool that was developed as part of the quality management movement.

It is a very valuable tool that can be used alongside the 5 Whys to break down the causes of a problem into a number of specific categories, and narrow down to get to possible root causes. These categories are: Man (or people), Machine (or equipment), Method (or process), Materials and Environment. These categories are the core elements of your farm and all problems will generally be a result of one of these elements.

A fishbone diagram (figure 9.8) provides a basic structure, and ensures that all possible causes of a problem are considered thoroughly, rather than just jumping to conclusions. It also encourages teamwork and open-minded brainstorming to develop the individual branches within each category.

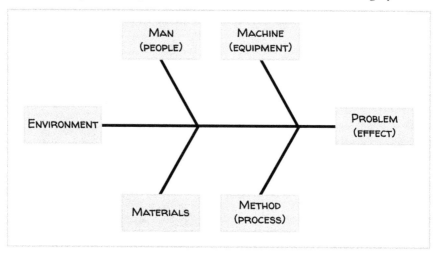

Figure 9.8: a fishbone diagram template

A Lean Farm example

Fishbone to the rescue

One farmer had a coliform and thermoduric grade problem on her farm after several years of no grades. She went home after learning about the fishbone problem-solving tool and spent the evening discussing and brainstorming the problem with her husband using a fishbone diagram. The next day the farmer presented the start of their fishbone. This approach meant that they considered all possible elements of the problem and could work through the problem in a thorough, structured way rather than making assumptions.

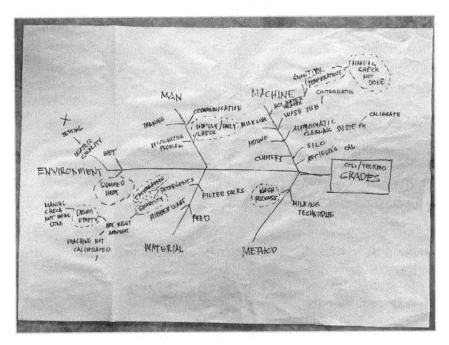

Using a fishbone diagram to investigate the causes of coliform and thermo-duric grades

To complete a fishbone, start by considering all the possible causes. Keep your brainstorming wide and open so that you consider absolutely everything that could be a cause of your problem. This eliminates the potential of ignoring an important contributor by just jumping to what you think is the cause. This is then followed by using facts and data to narrow down the potential causes. You cross out all unlikely causes until you only have the valid few, which you investigate further. Here are the steps to follow.

Step 1: agree on the problem

Decide with the team what the problem statement will be. Write this in a box on the right side of the fishbone diagram.

Step 2: brainstorm possible causes

For each category (e.g. Man, Machine, Method), ask the team to brainstorm what elements under that specific heading could have been a cause of the problem being investigated.

Step 3: brainstorm next-level causes

Once the first level of potential causes is identified under each heading, you can brainstorm the next level of potential causes, allowing you to branch out for each heading even further.

Step 4: repeat, branching out

Continue to brainstorm, spreading out the branches under each key heading (you can use the 5 Whys for this) until you can't go further.

Step 5: data-based elimination

Collect data and facts where possible and discuss each potential cause based on these data and facts. Eliminate any causes that can't be actual causes based on what your data and facts are telling you.

Step 6: root causes

Once you have eliminated all the impossible causes, you should be left with two or three root causes. These can now drive your improvement actions.

Step 7: monitor

Once you have implemented some countermeasures to fix the root causes, always collect data and monitor the improvement to ensure that the problem has been solved. You can use the PDCA cycle introduced in chapter 7 for this.

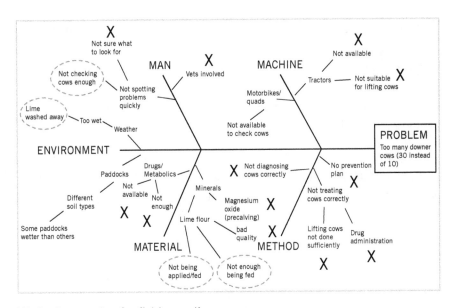

Worked example of a fishbone diagram

Lean Farm team activity

Using a fishbone diagram

1. Ask the team to think of a problem you are currently experiencing that you have been discussing for a while and is still not fixed.

2. Use a fishbone diagram to do a brainstorm of all the possible causes of the problem.

3. Discuss each cause and use fact-based elimination to narrow down the possible causes.

Summary

Practical problem solving is very important for every farm. Using simple problem-solving tools to identify the root cause of problems can help you eliminate a lot of waste on your farm. Much of our time in a day is spent fixing problems that are repeated and haven't been fixed correctly before.

Solving problems properly the first time using structured, thorough thinking and teamwork will save you time, money and frustration. Every person on your farm should be encouraged to use the 5 Whys every day. When a problem occurs that requires solving, don't spend your team meeting having a general discussion about it with people jumping to ideas such as 'it could be the wiring in the box', or 'maybe the filter is blocked', and then leaving the meeting with people going to try out a number of things. Two weeks later the problem is still there and the same discussion happens all over again. Instead, set a 30-minute to one-hour time limit, sit everyone down and work through a structured fishbone diagram so that the team can systematically work through the problem, identify all causes and narrow the causes down using facts and logic.

The above scenario has happened plenty of times on our farm while I cringed and then finally got the fishbone out!

A good idea is to set up a permanent fishbone template on a whiteboard in your team room. This way it is always on hand for the team to use whenever a problem arises.

Lean Farm team quiz:
PRACTICAL PROBLEM SOLVING

1. Who should solve problems?

2. What is a problem?

3. What is the difference between a Level 1 and Level 2 problem?

4. What is the 5 Whys?

5. How do you draw a fishbone diagram?

6. What do you use a 'problem and countermeasure' sheet for?

7. Why should problems be identified immediately?

Lean Farm action plan:
PRACTICAL PROBLEM SOLVING

These actions are included to provide some simple guidance for your farm. They are aimed at giving you a little bit of motivation and direction if you need it. Of course, the idea is for you to do as much as you can to gain the maximum benefit of practical problem solving on your farm.

1. Introduce practical problem solving to your team.

2. Identify at least two recent problems on your farm.

3. Do a 5 Whys on these two problems with the team to identify the root cause.

4. Identify a challenging problem that you currently have which you have not been able to solve yet.

5. With the team, use a fishbone diagram to solve the problem by doing a wide, open brainstorm and then narrowing down the causes.

6. Introduce a 'problem and countermeasure' sheet on your farm and get the team to start recording daily problems.

7. Discuss the problem and countermeasure sheet at each team meeting to follow up on actions.

8. Develop a permanent fishbone template on a whiteboard to have in your team office so that you can use it anytime.

9. Take *before* and *after* photos where possible.

You and your team are now equipped with some very useful tools to help you solve any type of problem effectively. Can you imagine the amount of time you will be saving each day when you don't have to keep fixing repeat problems? Next up we will talk about a more cultural element of Lean: built-in quality.

Chapter 10

Built-in quality

Quality is another fundamental element of a Lean business. Quality is more than just getting no milk quality grades. It is a fundamental part of a business culture.

In Toyota, quality is one of the pillars of the Toyota Production System. It is called *Jidoka* in Japanese. *Jidoka* is the philosophy of 'building in quality' to a process. This means delivering acceptable quality that meets the customer's expectation at every step in the process. The end result is that you are ensuring a good-quality product or service is produced and delivered to the customer from start to finish, not just with a final quality check.

Importantly, the philosophy of *Jidoka* provides the ability and expectation that the process will stop if a quality problem or abnormality occurs. By stopping the process, a defect or error can be fixed immediately, ensuring that the product or service is in an acceptable condition before it moves on to the next step in the process. At Toyota we used to define *Jidoka* as 'automation with a human touch' or 'autonomation'. This is a vital part of Lean and one of the core ways of operating in a Lean business. See figure 10.1 (overleaf) for a visual guide to the Lean vs traditional approach to quality.

DO NOT PASS A DEFECT ON TO THE NEXT PROCESS!

By adhering to the principle of *Jidoka* you are ensuring that you are building quality into your process at every stage. This requires a significant cultural shift as it makes *everyone* responsible for their quality. The expectation is that the work every person does is 100 per cent correct before it is passed on to the next process or person. It also means that your 'internal' customers (in other words, your colleagues) are just as important as your external customers when it comes to quality. You should deliver a high-quality product or service to your internal customers, not just to your external customers.

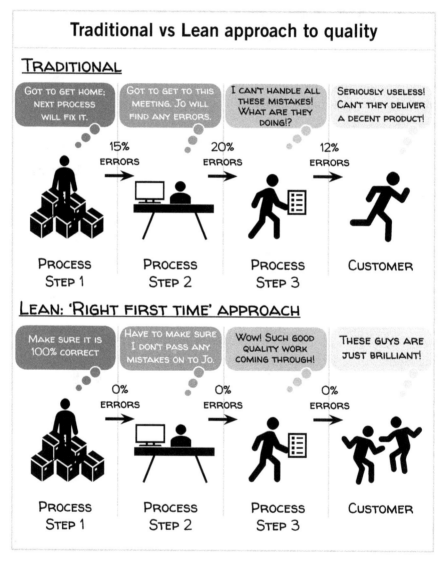

Figure 10.1: traditional vs Lean approach to quality

There are a number of systems you can use to stop the process when a problem or error occurs. The systems can be either manual or automatic. Automatic systems are more fail safe. A combination of both systems is commonly used.

Lean Farm team activity

Right first time

With your team, discuss the following:

— Do we do our jobs 100 per cent right first time?

— What examples can you think of where a job was not done right the first time?

— What was the impact of not doing it right the first time?

What is quality?

Quality may have a number of definitions for a number of people. A simple way to look at quality is to ask if the process or job is:

- *100 per cent right first time* (zero defects/errors)
- *100 per cent repeatable* (everyone)
- *100 per cent reliable* (every time).

Zero defects requires every person doing what they agreed to do or should do, when it needs to be done. It means that requirements and expectations must be clear (for example, having clear standards), training is adequate and relevant and everyone has a positive attitude.

Importantly, quality is based on *prevention* not *detection*. While *Jidoka* ensures you are detecting any quality problems and resolving them immediately, the focus of building a quality culture should be on developing such good processes that the problems are prevented from occurring in the first place. This is true, reliable quality. It requires all the elements of Lean coming together: proactively searching for problems; if any quality issues arise, immediately stopping them; finding the root cause and eliminating re-occurrence; and standardising to prevent errors. Prevention is much harder than detection—it relies on changing people's behaviours and beliefs.

QUALITY COMES FROM PREVENTION, NOT DETECTION!

If you don't focus on prevention, your business and team will forever be hoping like mad that you have managed to detect a problem and stop it being passed on to the customer. This isn't a very sustainable model.

Why should we strive for zero defects?

Our farm processes are very much interrelated. In everything we do, we have to ensure that we do it right, so that the dependent processes are not negatively affected.

All of our processes have variation associated with them unless we have excellent standardisation or automation that guarantees the same result every time. This variation leads to defects, errors or mistakes in the process. Defects or errors in one process can lead to defects, errors or problems in another process further downstream. As defects flow downstream, they usually escalate and so does the cost of these defects.

Focusing on quality and achieving zero defects will help our farms to:

- eliminate waste
- reduce cost
- improve productivity
- reduce time
- reduce frustration.

Jaya and Rob tagging cows—it is critical to get this process right

Our interrelated farm processes

Many of our on-farm processes are dependent on each other. If we do a bad job of raising our young stock, they come into the main herd underweight and go on to have calving problems, diseases and potentially become poor milkers.

Similarly, if we do a poor job of our pasture management, our cows don't eat enough, they lose condition, don't produce enough milk and our milk production targets are not achieved. You can see how doing the right thing at the right time can dramatically impact your overall business results and processes. Everything that someone does, no matter how trivial it may seem, will have an impact on another process, job or team member. We must aim to always do it right first time.

The cost of poor quality

What happens if we have an error and pass it on? As I described in the previous section, letting an error pass down a process rather than stopping it as soon as it occurs can have very negative effects on your farm's business, as you can see in figure 10.2.

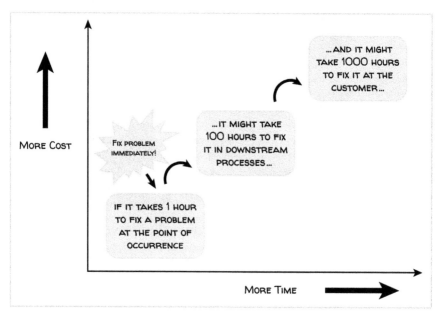

Figure 10.2: poor quality becomes more costly if passed on

Not only can errors or defects generate considerable costs to your business and consume time, but they can also have a negative impact on your farm image or brand if trust is lost.

Would you be happy if things were done correctly only 99.9 per cent of the time? Do you think this is good enough? Would you accept it? Do you know that if things were done 99.9 per cent right there would still be 113 unsafe plane landings across the world every 24 hours? Now, is that acceptable? Would you be confident taking a flight? Probably not, right?

QUALITY MATTERS! RIGHT FIRST TIME, EVERY TIME!

A Lean Farm example

When a defect escalates

Imagine if one day a bunch of cows that had antibiotics in their system got mixed up with the main herd. What could the impact of this mistake be?

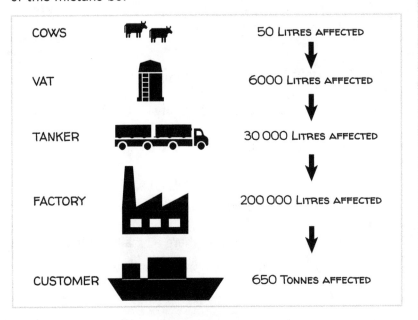

COWS		50 LITRES AFFECTED
VAT		6000 LITRES AFFECTED
TANKER		30 000 LITRES AFFECTED
FACTORY		200 000 LITRES AFFECTED
CUSTOMER		650 TONNES AFFECTED

You can clearly see how a small problem that could be easily managed on the farm can suddenly escalate into a very serious and significant problem.

Misconceptions

Many people think it is impossible to achieve 100 per cent excellence in quality every time. That errors are inevitable in any process. Do you think this is true? Some people believe employees would need constant monitoring and that they can't be trusted to get it right every time.

I believe that if you create a culture in which the people feel valued, empowered and included, people will naturally strive to do their best.

There may be circumstances where errors do occur because we are human; however, even if they do they should not be ignored or tolerated. Immediate action should take place to eliminate future errors from the same source.

Lean Farm team activity

Quality culture on the farm

As a team, discuss the following:

— How important is quality on your farm?

— Do you think you have a good quality culture?

— Do people regularly pass on defects to another person/process?

— What kind of challenges do you think you have on your farm that are preventing you from having a quality mindset?

— Can you think of an example where a small problem could have led to a large expensive problem if it wasn't caught in time?

— What is one action each person can do to improve your farm's quality culture?

Building a quality culture on your farm

If every farm had a quality culture, a lot of the frustrations we deal with daily wouldn't exist. Our businesses would be reliable and efficient and our teams less stressed. To start to create a culture where everyone in the team cares about doing their job 100 per cent right first time, the business needs to live and breathe quality.

Your farm needs to first be aware of its current mindset and status around quality. It then needs to make quality a priority, a value and an integral part of the farm. Some specific actions you can take on your farm to begin to instil a quality culture include:

- starting to capture information about and create awareness of frequency of errors and defects and the *cost* of these in terms of time, resources, materials, money, customer satisfaction and animals
- recognising that problems are opportunities to improve and that improvement is everyone's job
- reinforcing that it is better for the team to be spending time on value-added activities rather than fixing defects or errors
- providing positive encouragement and recognition when people identify and raise problems or mistakes
- always focusing on the problem and not the person: your farm will improve when the errors and defects are eliminated for good
- not blaming: this will have the opposite effect and scare people from raising issues
- shifting mindsets from 'we tolerate errors' to 'we have *zero tolerance* for errors'
- starting to think about prevention instead of just 'detection and fixing'.

The Lean approach to managing quality problems versus the traditional approach can be seen in figure 10.3 (overleaf).

Errors and defects

It is important to make the distinction between errors and defects.

An *error* is something that causes a process, service or product to deviate from the expectation or standard. Often errors are not easily visible.

A *defect* is a product or service that does not meet the minimum set requirements or standards, whether they be internal farm requirements or the customer's requirements. Defects are the result of errors made. However, not all errors result in a defective product or service. Sometimes you may get away with errors. Defects are tangible and visible. There may be multiple errors that lead to a defect.

Figure 10.3: traditional vs Lean approach to managing quality problems

DEFECTS ARE CAUSED BY ERRORS!

Examples of what can cause errors on the farm include:

- *no or inadequate processes*—there is no process in place to explain how to do a job correctly to meet requirements, leaving it to speculation
- *processes not being followed*—people ignore the process, don't do it correctly or forget key steps in the process
- *setting-up errors*—setting up equipment or plant incorrectly or using the wrong settings or tools
- *inadequate equipment or tools*—trying to do the job without the right tools or equipment and improvising
- *equipment maintenance/repair*—tools or equipment are poorly maintained or repaired, resulting in issues
- *unclear communication*—requirements not communicated clearly can lead to confusion and guesswork.

It can be a challenge doing things right the first time during calving

A Lean Farm example

That's not my calf!

We all know how difficult the calving period is. Trying to identify calves born during the night and recording their data at 4 am when it is dark and raining can be a challenge. I have seen many farmers try to do this in the pens later. Unfortunately, this can cause problems. If you don't tag and record a calf straight away in the paddock, the chance of errors in matching the right calf with the right cow and recording the correct calving data increases. These errors can then lead to defects such as the wrong calves added to the herd, or sold.

We are only human

As humans we are prone to making errors. These errors result in defects. There are multiple reasons why we make errors—sometimes we are at fault due to carelessness and other times they are genuine mistakes. Errors can be the result of many circumstances, such as:

- *new employees*—not familiar with the work environment, equipment or processes
- *poor focus/concentration*—a team member forgets what they are doing or misses steps such as sending wash water into the Vat of milk after forgetting to close taps
- *poor standardisation*—no clear processes are in place to show people how to do something
- *fatigue*—this is very applicable to farming as people are tired and easily lose concentration or do things in auto mode
- *intentional errors*—an employee thinks their way of doing a task is better and ignores the standard (if it exists) without understanding the process requirements.

So how can we eliminate all these human errors? We can use a concept called 'poka yoke', or error proofing, to help stop our mistakes before we actually make them—in other words to *prevent* our errors from happening in the first place.

Lean Farm team activity

Human error

With your team, discuss the following:

1. Identify at least three mistakes that were a result of human error.

2. Why do you think each mistake was made?

3. Could the mistake have been prevented? How?

'Poka yoke' (or error proofing)

'Poka yoke' is a Japanese term used in the Toyota Production System as part of the *Jidoka* philosophy. It is derived from the words *poka* (meaning 'unplanned errors'), and *yokeru* (to avoid) and essentially means error proofing.

Error proofing is the process of predicting, detecting and preventing errors in a process that can affect the farm or customer through the use of preventative or warning devices, mechanisms or systems. These can be electrical or mechanical. Error proofing is fundamental to building quality into your processes and stopping problems from occurring. Error proofing helps our farms and people by:

- making it easier for people to avoid mistakes
- eliminating the need to check
- eliminating the need to remember every process step or criteria
- eliminating the need for multiple checks and controls
- eliminating the possibility of confusion or mix-ups
- ensuring forced adherence to standard processes and systems.

Farmers are generally naturally good at error proofing and have probably been doing it for years without realising that it is error proofing. Our farms are full of examples of error proofing.

The best types of error proofing devices are simple, inexpensive and automatic. The error proofing mechanism should be built into the process and placed as close to the potential point of error as possible.

Common examples of error proofing in our daily lives include:

- different sized nozzles on diesel and petrol pumps
- seatbelt warning lights and buzzers in your car
- springs for batteries in devices to prevent incorrect insertion
- sink or bath overflow outlets
- automatic pop-up in toasters
- pressure release valves

- microwaves only starting once the door is closed
- child-resistant tops on medicines and chemicals
- airport trolleys stopping when handles are released
- circuit breakers preventing overloads
- the autocorrect function in Word.

Quality means doing it right when no one is looking.

Henry Ford

Farm examples of error proofing include:

- padlocks on vat taps to prevent the incorrect milk being taken
- alarm sensors on vat taps that signal if the taps are not connected correctly
- temperature alarms on vats
- text alerts to phones
- effluent pond floaters to inform of levels
- sensors to alert when the teat spray level is down
- backing gate automatically stopping when it senses a cow
- platform bail displays flashing if there is a problem cow
- tractor maintenance light signals when maintenance is due
- ball cocks on cooler tank to show levels
- electronic sensors on effluent irrigators
- PKE vibrators to stop sticking.

Error proofing systems

There are two key types of error proofing systems that are used: warnings and controls.

Warnings

These are visual or audible signals or alarms that alert a person when there is some sort of problem or error. The warning will not prevent a problem from occurring. Examples are the various warnings in our cars or lights flicking on our control panels in the dairy shed.

Controls

These are mechanisms or devices that will prevent an error from occurring by physically not allowing it. This could be a process automatically stopping, not starting at all or not allowing a task to be physically done. It requires human intervention for the process to continue. A control ensures zero defects as it takes the human element out.

Andon

The Andon is another form of building quality into our processes. The term is a Lean term used by Toyota daily that refers to the ability of anyone and everyone to either call for help or stop a process immediately if they detect an error or defect. This enables the team member to address the problem straight away within their process instead of passing the defect on to the next process. By stopping the process, every team member is empowered and responsible for building in quality.

The Andon in the car industry is usually a physical cord that is pulled by a team member to stop the production line when an issue is identified. There could be a number of reasons why a person would pull the Andon cord including safety, machine issues, delays and of course errors or defects.

Any dairy farm that has a rotary dairy shed will also usually have an Andon cord. It is very similar to the one you find on car assembly lines and serves the same purpose.

We have Andon cords in our sheds above our platforms—the blue and red cords that are hung around the outside of the rotary platform above the bails—so if there is a problem, the team member can pull the red cord and the platform immediately stops. This way the team member can address an issue such as animal treatments immediately rather than letting the problem continue.

The Andon cord in our rotary shed

Lean Farm team activity

Error proofing on your farm

Could error proofing be of benefit to your farm?

— Walk around the farm and find examples of existing error proofing (*poka yoke*). List some of these examples.

— Identify some opportunities for error proofing on your farm.

— Think about some common or recent farm problems or mistakes. Could these have been prevented with error proofing?

Summary

A quality culture can truly transform your farm into a productive, high-quality, reliable business. It will also develop a team environment that is empowered, motivated, caring and respectful.

Having a quality culture on your farm means everyone strives to do the right thing, the right way first time and expects zero defects or mistakes. It is a mindset and a part of the DNA of the farm. Management needs to lead by example and constantly reinforce and talk about quality to start to drive the right behaviours and change the culture to be quality conscious.

'Poka yoke' or error proofing can make a significant difference on your farm by making it fail safe, preventing unnecessary mistakes and problems from happening in the first place, saving your farm and team considerable money and time fixing expensive errors. Try to design your processes, equipment or tools so they can't be used incorrectly. Where possible, build in quality by developing and installing warning or control mechanisms. These are often not as expensive or difficult as you may think.

Lean Farm team quiz:
BUILT-IN QUALITY

1. What is a quality mindset?

2. What does right first time mean?

3. What does error proofing mean?

4. What are the two types of error proofing systems?

5. What is the difference between error and defect?

6. What does 'poka yoke' mean?

7. What is an Andon?

8. Who is responsible for quality?

Lean Farm action plan:
BUILT—IN QUALITY

These actions are included to provide some simple guidance for your farm. They are aimed at giving you a little bit of motivation and direction if you need it. Of course, the idea is for you to do as much as you can to gain the maximum benefit of built-in quality on your farm.

1. Introduce the concept of built-in quality to your team.

2. Identify all the errors/mistakes/problems that have happened on your farm over the past month.

3. Identify examples of error proofing you already have on your farm.

4. Discuss if the recent errors could be prevented using an error proofing mechanism.

5. Design and implement at least two error proofing devices that will stop two of your errors/problems/defects.

6. Monitor your error proofing mechanisms to ensure they have prevented the mistake or error from occurring again.

7. Take *before* and *after* photos of your processes/error proofing ideas.

You now have a standardised and high-quality farm that doesn't have errors, make mistakes or have defects. Isn't life so much easier? There is still another huge part of our farms that remains a constant source of frustration for many farmers—maintenance. Well Lean addresses this too. I want to now share with you something called Total Productive Maintenance.

Chapter 11

Total Productive Maintenance

Total Productive Maintenance (TPM) is not necessarily a Lean concept. Versions of this have been around for decades but it is a highly valuable concept that is particularly applicable to farming and should be instilled in every farm. TPM is essentially a holistic system for managing your farm equipment, machines, facilities and plant. It is an approach to maintenance that will help your farm to become a reliable, efficient and preventative farm, rather than being unreliable, ad hoc and reactive. Like quality, TPM requires a shift in mindset.

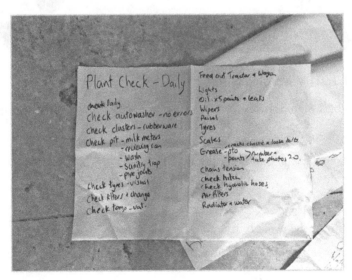

Brainstorming required preventative maintenance tasks

TPM is made up of several key elements. There are a number of different models for representing TPM and these elements. Figure 11.1 illustrates one version I have constructed that fits well with farming.

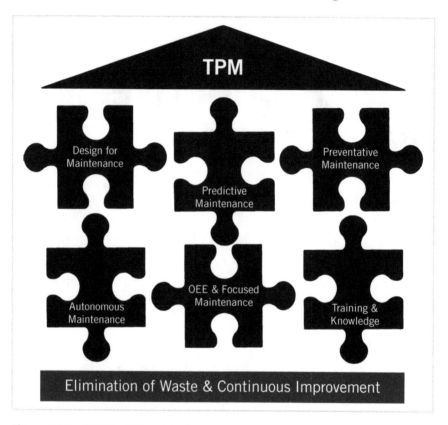

Figure 11.1: TPM building blocks

TPM can have significant benefits for your farm and team because it helps to:

- eliminate waste
- prevent unplanned breakdowns
- improve your understanding of your equipment, plant and facilities
- optimise the availability of your equipment and plant
- ensure your equipment and plant are operating efficiently and correctly
- react to any breakdowns effectively and quickly

- improve your team's knowledge of your plant and equipment
- plan repairs or maintenance when it suits you rather than at the last minute.

Farm vehicles

Probably one of the biggest frustrations on the farm is farm vehicles. Most farmers will agree that quads, rangers, bikes and tractors are constantly breaking down. If you started analysing the reasons for these breakdowns, most would be a result of poor maintenance. Either oil hasn't been filled up, the bikes haven't been cleaned sufficiently, if at all, or scheduled maintenance is well past the due date.

The result is that the vehicles break down in the middle of a paddock when you are in the middle of an important job and need them. This generates rework, wastes time and costs money. A proper maintenance program for all our farm vehicles can help to improve the condition of the vehicles, manage cost and prevent unplanned breakdowns.

Lean Farm team activity

Farm vehicle maintenance

Discuss the following with your team:

1. When was the last time a farm vehicle broke down? Why?

2. Who is currently responsible for maintaining the farm vehicles?

3. Who should be responsible?

4. How often do your farm vehicles get checked and cleaned by the team?

Imagine the time and money it would save us if we never had a broken-down bike or farm vehicle again. Imagine the frustration it would save us if our tractors were always clean, well maintained and greased, and available when we needed them. Would we be much happier? Would our farms operate more effectively? And imagine if we applied this to every other piece of equipment we have on farm: our irrigators, plant, effluent saucers, pumps and so on. With everything always working correctly like clockwork, we would have a pretty impressive farm!

Why do we have maintenance concerns?

Every farmer would probably agree that maintenance is a big part of their job and also a key frustration. Maintenance is particularly important on a farm where equipment is usually exposed to weather, mud and animals. There are many reasons why equipment, plant or facilities break down, such as:

- *failure to maintain machines regularly* — for example, required servicing; maintaining oil levels; cleaning; tightening, checking, replacing worn components
- *failure to maintain correct operating specifications* — for example, ensuring equipment settings such as speed, pressure and temperature are correct
- *incorrect use* — for example, the team isn't trained on how to use the equipment properly or are careless and misuse equipment
- *inadequate design* — the equipment hasn't been designed well, resulting in difficult access and use, and poor functionality.

Farm examples of equipment/facilities that need TPM include:

- irrigators
- irrigator sheds and pumps
- effluent ponds
- effluent spreaders/pods
- pumps in general
- water tanks
- Vats
- shed production plants
- shed rubberware/liners/ milk filters
- troughs and water piping
- fence electricity
- bores
- calving equipment — cafeterias, milk warmers
- farm vehicles
- trailers
- buildings
- team houses
- backing gates

- post rammers
- power tools

- workshop facilities
- implements.

Keeping a record

The best way to get a handle on how much maintenance your team does and what type of maintenance you need to do, is to track it. Most farms I have seen don't do this very well (other than maintaining vehicles) and then wonder where all their time is spent or why the maintenance costs are so high.

Tracking every maintenance task for a week or two will help your farm to be in a better position to understand where you are having issues with maintenance and how big the issue is. You can then target your improvements or TPM activities on these problem areas. It will also tell you if there are repeat problems or if a problem has in fact been fixed after an expensive visit from the electrician.

Two of our team's sparkling clean bikes

A Lean Farm example

Track your data

Last year we decided to track our maintenance activities. We discovered that there was a lot more maintenance being done than we expected. It was also surprising how much time the team was spending on breakdowns and other maintenance issues. We found that there were many issues where suppliers had to come back several times because they weren't fixed the first time. This obviously significantly added to the cost.

By analysing all of the data collected we were able to target our improvements on priorities and do structured problem solving to identify root causes and develop suitable solutions. We could also quickly estimate the cost of these maintenance issues to the business. By recording the data on a maintenance record sheet—which lists the date, the affected equipment, the impact the problem had, the downtime it created, and how and when the problem was rectified—we were able to determine that one repeat issue was a malfunctioning teat spray. We had the supplier attend to the problem twice, resulting in 12 hours of downtime. Fortunately it wasn't in the middle of calving where 12 hours is hard to find.

Tracking data like this helps you to very quickly see where you have reliability issues with your equipment, why the breakdowns occurred and how much disruption and cost it caused. It also gives a very good indication of where time is being spent when you are wondering why some of the more value-adding jobs aren't getting done.

Figure 11.2 is a worked example of a maintenance record sheet that you can use on your farm.

MAINTENANCE RECORD SHEET

#	DATE	RAISED BY	EQUIPMENT/ FACILITY AFFECTED	PROBLEM DESCRIPTION	IMPACT TO FARM/ BUSINESS	DOWN-TIME	SUPPLIER/ CONTRACTOR INVOLVED	ACTION/COUNTERMEASURE	ANY ISSUES/ COMMENTS	DATE FIXED	WHO
1	8/11	JS	PLATFORM	WATER LEAK UNDER PLATFORM	WATER LOSS	NA	PUMP GUYS	CUT & BLOCK PIPES AND RUN NEW PIPES OVER TOP	ADD DRAINS	8/11	JS
2	10/11	RK	QUAD 10	RATTLING UNDERNEATH	1 QUAD DOWN	1 DAY	MOTOR BIKE GUYS	REPLACE GEAR BOX		11/11	RK
3	15/11	RC	AFI	NOT RECORDING CERTAIN BAILS	MISSED COW DATA	6 HRS	PLANT GUYS	CLEAN SCANNERS, RESET CONTROL PANELS		20/11	SC
4	17/11	JW	MEAL FEEDER	INCORRECT AMOUNT OF MEAL/ MINERAL BEING FED	PROD LOSS	3 DAYS	ELECTRICIANS	WIRING INCORRECT—WIRE CORRECT CONTROL BUTTON TO CORRECT FEED	REWORK—NOT INSTALLED RIGHT	18/11	JS
5	18/11	HB	IRRIGATOR	HOSE LEAKING							

Figure 11.2: a worked example of a maintenance record sheet

The pillars of TPM

Each of the elements of TPM are important to ensure a thorough system for maintenance. The best situation would be if we didn't need to do any maintenance at all. However, this is unlikely as this is usually reliant on the design of our equipment and therefore not usually within our control. What is in our control is how we manage our maintenance and our equipment in our business. We can either make it easy and efficient, or do nothing and let it be a constant source of pain for us.

On our farm we are just starting to use TPM, but we have a long way to go. We most certainly have seen and dealt with many problems on a daily basis that can easily be prevented. TPM is one of our key focus areas for the next year.

Farm examples of maintenance issues that could be prevented include:

- tractors catching alight because of bird nests
- feed hoppers breaking down because they contain bird nests
- pumps with blocked filters
- pumps not working because they need greasing
- blocked teat sprays
- plant stopping due to a lack of oil
- tyres bursting because of overloaded mixer wagons
- the Vat temperature gauge failing
- milk residue building up on the plant
- rubberware splitting
- motor bikes needing oil
- pasture readers snapping off after taking a corner too quickly
- control panels not working
- broken springs on cup removers.

Design for maintenance

If we consider maintenance and usability at the design stage of any of our plants, equipment, vehicles and facilities it helps us to build things that make usability and our maintenance activities easy. I often wonder if someone has actually tested the usability of equipment before selling it on to customers. It also appears to me that many designers and manufacturers seem to expect that the equipment will never break down. It is often not designed for maintenance, making it very difficult to access key areas to perform any required maintenance.

A Lean Farm example

No easy maintenance

One of the first things I noticed when I came to the dairy farm was how poor design made it difficult to do basic cleaning and maintenance on the shed equipment. For example:

— The rotary platform inclines upwards with a rim along the inside edge so that when you try to hose it (using the hoses that are placed around the outside perimeter of the platform) you are hosing all the dirt upwards instead of downwards with the natural flow of gravity and it all gets stuck up against the lip. You then have to try to hose on an angle to get all the dirt out so it can flow into the drains that are positioned on the ground.

— When doing a daily plant wash, team members need to reach, turn, bend, stretch and step in awkward positions and corners to turn various taps and changeover hoses. Further water runs out of hot water tank taps onto the shed floor before running to a drain positioned away from the tap.

Mat hosing down the platform

While most farmers are not in a position to be directly involved in the design of key equipment, if you do happen to be working with a company that is building or designing something for you, try to get involved early on. Ideally, anyone in your team who will need to use the specific piece of equipment should be involved in the design and specifications and generally have an input into the buying decision-making process.

This will ensure that the equipment bought is fit for its purpose, reliable, and easy to use, clean and maintain. Don't just go and buy a piece of equipment without thoroughly checking it for design flaws that make it difficult to maintain, operate and clean. Things to consider when designing new equipment or facilities or making buying decisions are:

- What kind of maintenance is expected? How often?
- Who does this maintenance?
- How easy is it to do the maintenance?
- How well can you access key parts of the equipment?
- Is it easy to clean if needed?
- What training will the supplier provide?

- How much after-sales support is there?
- How easy is it to get spare parts? What is the cost? How long will it take to get them?
- How easy is interchangeability?
- Who owns the relevant data?
- What support documentation/manuals are provided?

Figure 11.3 (overleaf), gives you a run-down of all the considerations to make before committing to new equipment/facilities.

Don't just buy equipment — question everything about it!

Lean Farm team activity

Unplanned maintenance on the farm

With your team, discuss the following.

1. What recent maintenance issues have you had on the farm?

2. Were they unplanned?

3. Could they have been prevented?

4. What are some examples of equipment/facilities that could benefit from a maintenance program?

5. What examples do you have on the farm where equipment is not designed for easy maintenance, cleaning or usability?

6. How do you currently record or track maintenance?

7. How much time in a day do you spend dealing with unplanned breakdowns or maintenance?

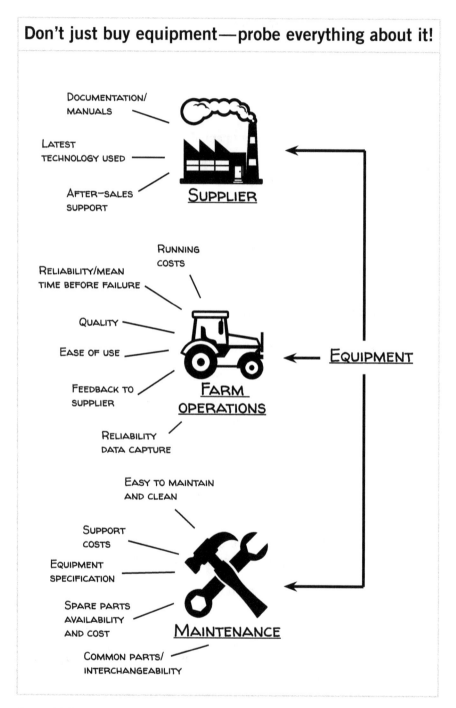

Figure 11.3: understand and probe your equipment before buying it

Predictive maintenance

Predictive maintenance involves predicting when a piece of equipment or facility is going to fail and being able to do the required maintenance or repair just in time, prior to a failure. It is essentially a forecast of potential problems and then planning for them.

If we understand our equipment's life cycle and life span, we can estimate when maintenance tasks might be required and plan for these. There may be certain signals or signs that are either built into the equipment or we can easily identify when something is about to break down. For example, our water filter starts beeping when the UV light bulb is coming to its end date, informing us that we need to replace it before it is no longer effective rather than the whole filter just stopping.

These days there are a variety of methods, including artificial intelligence, being developed to automatically diagnose and predict whether a machine is going to break down by monitoring sounds, frequencies, vibrations and other parameters. One day we will be able to save ourselves a lot of time in planned maintenance activities by simply waiting for our machines to tell us when they are going to break down and what maintenance activity they need from us!

Our farms and sheds are full of facilities and equipment that need maintenance

In the meantime, you can think about whether some of your equipment at least gives you telltale signs of potential failure before it happens. Then you can build this into your productive maintenance system.

Planned preventative maintenance

This the most commonly known element of TPM. It involves identifying the types of maintenance activities required for all of our equipment and facilities and how often these are required to prevent any unplanned breakdowns. A plan or schedule is developed to complete the required maintenance activities at the right time. Planned preventative maintenance allows our farms to monitor, maintain or repair equipment before it fails and causes an unplanned stoppage.

On our farm, we are working on a standard whole farm preventative maintenance schedule that will be visualised on a large board at each farm and monitored weekly during the team meeting.

It means we can plan ahead, know what equipment is due for servicing or maintenance, and schedule our farm activities accordingly. For example, we won't ever have a situation where all of our tractors go out to get their annual service at the same time in the middle of calving.

Here are some key steps to help you establish your preventative maintenance program:

1. Identify the categories of equipment that you need to maintain regularly — this could be the shed, vehicles or auxiliary equipment.
2. For each category, identify the specific equipment, facilities or items that need maintenance.
3. For each piece of equipment or machine, identify all the parts that need to be checked, maintained, cleaned, serviced or repaired.
4. Establish how often the maintenance task is required — you can use supplier guidelines, manufacturer specifications, historical data of wear and tear or breakdowns.
5. Determine who should be responsible for each task — is it an internal or external maintenance task?

6. Develop visual standard work for each maintenance task to explain how it should be performed correctly. Include how long the task should take.

7. Develop a preventative maintenance schedule or matrix listing all the planned maintenance activities.

8. Establish a spare parts strategy that suits your maintenance program.

9. Establish a clear training plan if the team needs upskilling to be able to complete the maintenance activities.

10. Continuously improve and evolve the preventative maintenance schedule including adjusting maintenance task frequencies as needed.

We have an annual preventative maintenance plan that we have started documenting for our farm. It is in a check sheet–type format so that once the scheduled check is completed it can be signed off by the person. This way we can immediately see if we have done all the checks required for that week or what checks are coming up. At this stage, ours is only a draft and far from comprehensive, but once the draft is completed and we are satisfied, it will be printed on whiteboards and hung on a wall as a working visual preventative maintenance board. You could also create a weekly version of this maintenance plan.

Lean Farm team activity

TPM farm observations

With the team, walk around the farm and identify examples of good maintenance activities and planning (e.g. preventative maintenance).

— Are team members involved in maintenance activities?

— How is maintenance being recorded and tracked?

— Is there a preventative maintenance schedule?

— What kind of checklists exist for maintenance activities?

Figures 11.4 and 11.5 show examples of the templates we are developing. You can obtain these online at www.leanfarm.nz if you don't want to recreate them.

ANNUAL FARM PREVENTATIVE MAINTENANCE SCHEDULE

EQUIP/FACILITY	#	MAINTENANCE TASK/CHECK	FREQUENCY	WHO	JUN	JUL	AUG	SEPT	OCT	NOV	DEC	JAN	FEB	MAR	APR	MAY
PLANT	1	RUBBERWARE CHANGED	2 × YEAR	JS	✗						✗					
	2	PLANT GENERAL CLEAN	2 × YEAR	JS	✗						✗					
	3	MILKING PLANT CLEAN	2 × YEAR	JS		✗						✗				
	4	PLANT CALIBRATION	2 × YEAR	JS		✗						✗				
	5	PLANT CHECK	MONTHLY	SC/JW	✗	✗	✗	✗	✗	✗	✗	✗	✗	✗	✗	✗
IRRIGATION	6	SERVICE IRRIGATOR	ANNUAL	JS						✗						
	7	CHECK PUMPS	ANNUAL	JS						✗						
VEHICLES	8	MAIN SERVICE	QUARTERLY	JS		✗			✗			✗			✗	
	9	THOROUGH CLEAN	MONTHLY	SC/JW	✗	✗	✗	✗	✗	✗	✗	✗	✗	✗	✗	✗
	–															

Figure 11.4: example of a preventative maintenance schedule

WEEKLY FARM PREVENTATIVE MAINTENANCE SCHEDULE

EQUIP/FACILITY	#	MAINTENANCE TASK/CHECK	FREQUENCY	WHO	MON	TUES	WED	THUR	FRI	SAT	SUN
PLANT	1	DAILY 5S	DAILY	RC	✖	✖	✖	✖	✖	✖	✖
	2	VAT GAUGES WORKING	WEEKLY	RC			✖				
	3	MILKING PLANT WORKING	WEEKLY	RC	✖						
	4	BAILS WORKING	WEEKLY	RC				✖			
	5	MEAL FEEDER WORKING	DAILY	RC	✖	✖	✖	✖	✖	✖	✖
	6	CHECK OIL/LIGHTS	WEEKLY	AR	✖					✖	
TRACTORS	7	CHECK BIRDS NEST	DAILY	AR	✖	✖	✖	✖	✖	✖	✖
	8	CLEAN	WEEKLY	AR					✖		
FARM BIKES	9	CHECK OIL/FUEL/LIGHTS	DAILY	SC	✖	✖	✖	✖	✖	✖	✖
	10	CLEAN	2 × WEEK	SC	✖				✖		
....	⋮										

Figure 11.5: example of a daily/weekly maintenance check worksheet

Spare parts

Have you ever had to do a last-minute run to your plant parts supplier to buy a seal or special component so that you can quickly get back to the shed and fix a piece of equipment? This has happened many times on our farm. But how do you know what spare parts you should keep? Unless you have a dedicated warehouse, it is not practical or cost effective to keep a spare of absolutely everything. If you collect data on your maintenance activities, you can do a reasonable logical analysis of what spares are worth having. Here are some key points to consider:

- Weigh the impacts of disruption to your farm of no parts vs the cost of keeping spares.
- Identify high cost spares versus low cost spares.
- Know the lead times for critical spares and how easy it is to get them.
- Find local suppliers of spares where possible.
- Work out what storage space is required and where.
- How can spare parts be tracked to know their condition and whether they are calibrated/working correctly?
- What common parts are there so you can share one spare among multiple pieces of equipment?

A Lean Farm example

In among the blackberry bushes

There is no point keeping spare components or parts if they aren't taken care of. One farmer told me a story about a pump blowing up on a piece of equipment. The farmer spent several hours disassembling everything to get the pump out. He then asked if there was a spare pump anywhere and the farm owner told him there should be one 'in the workshop somewhere'. After searching the workshop, which was piled high with anything and everything and overgrown with blackberry bushes, the farmer finally located the ancient pump sitting on the ground under the blackberry bushes.

He then spent more time cleaning the pump and fitting it just to discover that this pump didn't work either.

This is a great example of poorly stored and maintained spares. It is a waste of time and money keeping spares if you can't find them or if they are just going to go rusty somewhere in the dirt. This isn't a spare-part strategy, nor is it conducive to preventative maintenance.

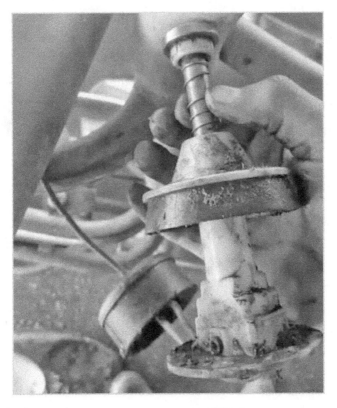

Fixing a broken spring on cup remover assembly

Lean Farm team activity

Farm preventative maintenance plan

Design a preventative maintenance plan for your farm.

1. Identify all the equipment, machines or facilities that require maintenance.

2. Identify the specific tasks required.

3. Determine how often the maintenance tasks or checks need to be done.

4. Who should do the maintenance?

5. Prepare a simple preventative maintenance plan document that visually schedules all your planned maintenance tasks.

Autonomous maintenance

Traditionally, maintenance is performed by 'maintenance specialists' or engineers. Some of the larger farms may have a maintenance technician who is dedicated to fixing equipment or machines. However, there are many maintenance tasks that can be done by everyone, every day.

Autonomous maintenance involves building maintenance tasks into a team member's daily job. This enables basic maintenance tasks to be integrated into our processes, performed continuously and be the responsibility of all team members. If regular maintenance tasks become a part of our standard processes, or our daily 5S, they are not seen as a separate, additional task. Furthermore, it ensures that our equipment and machines are looked after daily. By empowering team members to be responsible for basic maintenance tasks, we can develop their capability so that they can:

- spot inconsistencies or issues with equipment quickly
- perform basic tasks themselves without needing assistance
- troubleshoot more effectively
- understand equipment assembly and functionality
- prevent unplanned downtime
- take ownership of the equipment and lead maintenance activities.

The five key elements of autonomous maintenance are illustrated in figure 11.6.

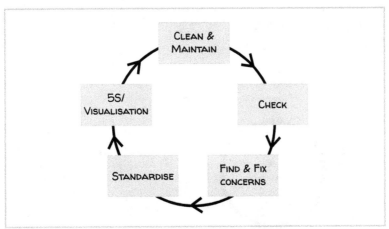

Figure 11.6: five key elements of autonomous maintenance

How to implement autonomous maintenance

Autonomous maintenance tasks should be built into the farm's everyday work. And many, such as motorbike checks and cleans, are already part of our standard. By identifying maintenance activities that need to be done hourly, daily or weekly, we can set up systems to ensure these get done and are not forgotten or missed.

One such system is called a Kamishibai (or T-card system). It is a simple visual way of knowing what needs to be done and tracking a required maintenance task to see whether it has been completed. You can set up Kamishibai boards on a wall so that everyone can see them and own them. Kamishibai systems can be used not just for maintenance activities, but also as a reminder of any regular activities or tasks, to complete necessary audits, for tracking projects or for quality or 5S checks.

Essentially, a Kamishibai board is a card rack (like the old clock-on, clock-off time-card racks). It has a number of cards slotted under each day, with each card describing a particular task that needs to be completed that day (see figure 11.7). Cards can be printed so that they are red on one side and green on the other. All cards start off displaying the red side, and when the task has been completed the team member can turn the card around so that it is green. This is a great visual way of seeing which tasks are outstanding.

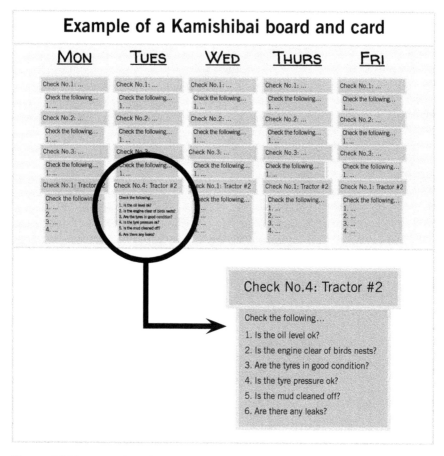

Figure 11.7: example of a Kamishibai card that gets slotted into the Kamishibai board

The key steps for setting up a Kamishibai board are to:

- identify the tasks that need to be completed hourly, daily or weekly
- record the description of each task on a card, including how long it should take, when it will be done and who is responsible

- allocate the tasks to days of the week, ensuring an even distribution of workload
- set up the card racks and place cards into the rack
- train all team members to be able to do every task on the rack
- monitor that the Kamishibai system is being used correctly.

This Kamishibai system can make incorporating simple, regular maintenance tasks into your team's daily activities very effective and easy. Your team will be able to simply have a look at the board and know immediately what action is required.

Lean Farm team activity

Farm Kamishibai

With your team, discuss the following:

1. What equipment or facility on your farm needs daily or weekly maintenance checks or tasks?

2. Write down all the maintenance tasks required for the equipment/facility and the frequency.

3. Design a T-card (Kamishibai) system for these maintenance items.

OEE and focused maintenance

The last type of maintenance is really about continuous improvement. By collecting data on your machine/equipment performance, you can identify where there are repeat problems, constant breakdowns or poor reliability. This can then help you to focus your improvement activities on getting to the root cause of these maintenance issues and fixing them for good, improving your overall farm operations. Your data collection and maintenance record sheet will be very useful for this.

What is OEE?

OEE (overall equipment effectiveness) is a vital standard Lean measure in manufacturing environments. It is a measure of our equipment or machine performance, and it identifies the percentage of time that our equipment is truly productive. OEE focuses on three key elements: availability, productivity and quality (see figure 11.8 for how OEE is calculated). All of these elements are affected by TPM.

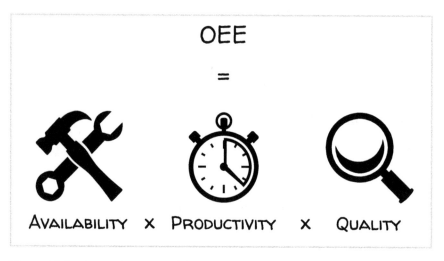

Figure 11.8: three elements of OEE

If you have any equipment or machinery that has a very low OEE figure, this suggests that the equipment needs significant improvement in performance.

How good are farmers at measuring the performance of their very expensive facilities or equipment? I don't know too many farmers who would know what OEE is, let alone use it as a measure. Yet, if you actually collected data on how often our dairy sheds are not running correctly you might discover that the OEE of the typical dairy plant is far from 100 per cent. Have a look at this example.

A Lean Farm example

The OEE of a typical rotary

Let's say we use our dairy shed for 7 hours a day, 7 days a week. That is 49 hours per week of available or planned production time. Out of that 49 hours, we have breakdowns amounting to 10 hours due to bails not working and the platform stopping. In addition, the platform is not operating at optimum speed as the cows would not get off it so it has to be slowed down continually. At optimum speed, the platform runs at 450 cows/hour. However, we have had to have it going at around 350 cows/hour. Finally, out of our 7 days of milking, the Vat taps weren't set up correctly on Tuesday afternoon and one full milking's worth of milk went down the drain. We have all probably experienced a week like this. So how does this affect our productivity? Let's work out the OEE of our dairy shed above.

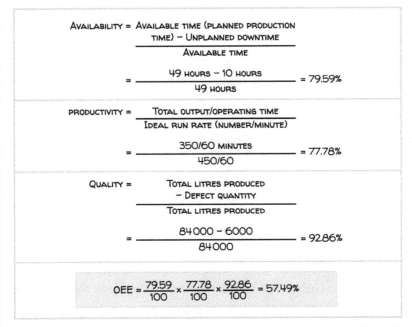

$$\text{AVAILABILITY} = \frac{\text{AVAILABLE TIME (PLANNED PRODUCTION TIME)} - \text{UNPLANNED DOWNTIME}}{\text{AVAILABLE TIME}}$$

$$= \frac{49 \text{ HOURS} - 10 \text{ HOURS}}{49 \text{ HOURS}} = 79.59\%$$

$$\text{PRODUCTIVITY} = \frac{\text{TOTAL OUTPUT/OPERATING TIME}}{\text{IDEAL RUN RATE (NUMBER/MINUTE)}}$$

$$= \frac{350/60 \text{ MINUTES}}{450/60} = 77.78\%$$

$$\text{QUALITY} = \frac{\text{TOTAL LITRES PRODUCED} - \text{DEFECT QUANTITY}}{\text{TOTAL LITRES PRODUCED}}$$

$$= \frac{84\,000 - 6000}{84\,000} = 92.86\%$$

$$\text{OEE} = \frac{79.59}{100} \times \frac{77.78}{100} \times \frac{92.86}{100} = 57.49\%$$

(continued)

A Lean Farm example *(cont'd)*

Basically what this OEE is telling us is that during this particular week our shed was only operating at 57 per cent efficiency during the mere 49 hours that it was in operation. This was because of our poor quality, unplanned breakdowns and sub-optimal speed. If we ignore the fact that our expensive asset is actually sitting and doing nothing for the majority of the week, should we be happy with this OEE? Is this the performance we expect from a shed that costs more than one million dollars?

In the manufacturing world, particularly in a Lean environment, this would be very unacceptable. The target should be 100 per cent — this means no errors, no unplanned breakdowns and optimal operating speed. At Toyota, I used to do audits of suppliers and looked at their OEE as a key metric of their capacity and capability. Often, anything below 90 per cent (particularly if that meant it affected their capacity to produce the volume required) was a red flag and usually meant the company had to undergo significant improvements. Even if their OEE was at 98 per cent and their capacity was at the limit, we identified opportunities to improve the OEE by looking at the components of quality, productivity and availability.

Training and knowledge

The final building block of TPM is your team's skill level. TPM won't be successful if your team isn't engaged and, importantly, doesn't have the skills to be able to complete the required maintenance tasks and plans. Your team needs to have a good understanding of their work environment including how key equipment and machinery works.

Developing standards, expectations and training plans will be a key component of introducing TPM on your farm successfully.

Summary

TPM is one of the key principles of Lean and it focuses on optimising our equipment and facilities. TPM can provide significant benefits to your farm. Our farms are full of machines, equipment and facilities that are prone to unplanned breakdowns. We spend a lot of time and money on maintenance activities and often take several attempts to fix an issue. Furthermore, when breakdowns occur randomly just when we need the piece of equipment, it leads to significant stress and frustration for our team and farm.

TPM can help you avoid unplanned breakdowns and maintenance work. It makes farms preventative, not reactive. As such, TPM can help you improve your efficiency, reduce waste, improve quality, save time and cost, and improve your team culture.

Lean Farm team quiz:
TOTAL PRODUCTIVE MAINTENANCE

1. What are the six elements of TPM?

2. What is the difference between preventative planned maintenance and predictive maintenance?

3. What is a Kamishibai system?

4. What is autonomous maintenance?

5. What does OEE stand for?

6. How is OEE calculated?

7. What does design for maintenance mean?

Lean Farm action plan:
TOTAL PRODUCTIVE MAINTENANCE

These actions are included to provide some simple guidance for your farm. They are aimed at giving you a little bit of motivation and direction if you need it. Of course, the idea is for you to do as much as you can to gain the maximum benefit of TPM on your farm.

1. Introduce the concept of TPM to your team.

2. Identify the key equipment/machines/facilities requiring maintenance.

3. Develop a Kamishibai system for regular maintenance activities or checks.

4. Develop an annual preventative maintenance plan.

5. Think about current issues with design of existing equipment.

6. How can your current farm equipment or facilities be better designed for maintenance?

7. Take *before* and *after* photos of your systems.

I hope that you are as excited as we are about introducing and using Total Productive Maintenance on your farm. I would love to envision a future when all our farms run smoothly, like a well-oiled machine, every day!

Flow is another core pillar of the Toyota Production System. Our next chapter discusses the effect of good flow on our farms.

Chapter 12

Creating flow on the farm

Another key pillar of Lean is something called just-in-time (JIT). This means producing and delivering JUST what is needed, JUST when it is needed and in JUST the amount needed. JIT production is based on customer demand.

By default, dairy farming isn't exactly just in time — it is more like a push model where we produce as much as we can, push it out of our farm gate, and get paid for it. We aren't asked to produce only the amount that the end customer actually wants. This model works well and is beneficial for us in times where there is endless consumer demand and we can't produce enough. However, in times where there is oversupply in the market and low demand, we end up with stockpiles of milk powder in warehouses and low milk prices as a result.

Despite this, we can still use the concept of JIT on the farm in our other work. JIT principles help us to make our processes flow better by reducing process lead time, and removing any stops, waits and other waste from the process. Trying to have one-piece flow in our process, rather than doing things in batches, will improve our process efficiency, quality and speed.

One-piece flow essentially means a FIFO system — first in first out. In other words, we process one and move one in the same sequence. How can this apply to farming?

There are probably two main examples of JIT and FIFO on farms:

- *robotic milking*—cows milk themselves Just when they need to and in Just the amount they want—perfect JIT

- *rotary vs herringbone dairy shed*—a rotary dairy shed is designed to create one-piece FIFO flow: one cow gets on the platform and moves around, and one gets off. The first cow that got on is the first cow that gets off (unless of course she is a long milker and goes around twice). In a herringbone, a batch of cows come into the shed and get milked, then that batch moves on and a new batch comes in. This is batch processing (see figure 12.1).

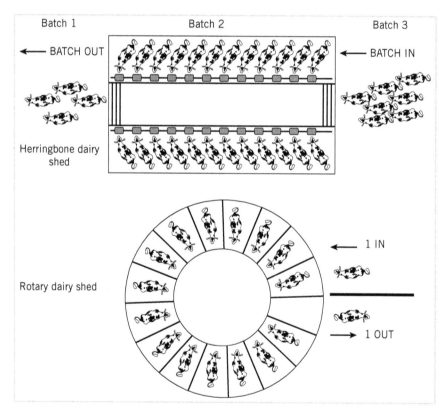

Figure 12.1: batch processing in a herringbone shed vs rotary shed

Grassmere rotary shed with its FIFO system

Lean Farm team activity

One-piece flow

With your team, discuss the following:

— Do you batch process or have one-piece flow in your farm processes? Give examples.

— What processes do you think could benefit from a one-piece flow model?

— For those who have worked in a herringbone shed and a rotary shed, which shed felt like there was better flow?

Creating a one-piece flow process can have significant benefits, including:

- *reduced process time*—it takes longer for a cow to be milked in a herringbone shed than a rotary shed because she has to wait for all the other cows in her batch to finish milking
- *improved quality*—if someone is doing a job incorrectly (e.g. tagging cows) and does a whole batch of cows before they get passed on to a second person who is recording tags and notices the mistake, they have created a lot more defects (a batch of cows) than if they had just tagged one cow and moved her on. The defect would have been identified immediately and corrected before more cows were tagged incorrectly
- *reduced waste*—batch processing can generate waste in additional effort, rework, waiting and processing too much too soon
- *less waiting time*—in a batch process, cows and people spend more time waiting. This is not only costly but also means that cows are standing on yards rather than eating grass. This ultimately reduces production potential and your profit
- *less stress*—when tasks and processes flow smoothly, the farm environment and team are less stressed. There are fewer stops and starts; less backwards and forwards; and more control.

Of course, we can't all run out and go and buy rotary sheds or robots. This needs to make overall financial sense for us. However, there are plenty of other examples on the farm where we can think about flow.

Spaghetti farm

Does it ever feel like you spend your day running around from one place to the next? If you drew a diagram of every one of your movements in a day, what do you think it would look like? Would it look like a bowl of spaghetti—in other words, something like figure 12.2?

If the flow of what we are doing, whether it is milking cows or fixing fences, looks messy then it probably is messy. Spaghetti flow creates inefficiency, frustration and chaos in our day. Flow can mean the flow of people, processes, cows, materials, machinery or information.

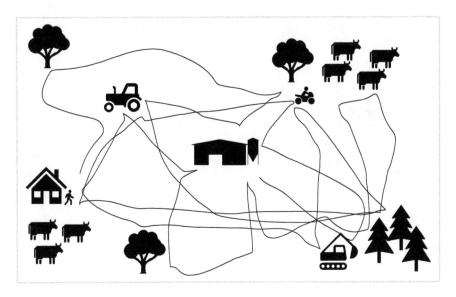

Figure 12.2: spaghetti flow on a farm

Spaghetti = chaos!

Flow = simplicity and efficiency

As farms get bigger, or older, or more competitive, processes get more complex, entangled or forgotten. Farms themselves also become more complicated as there are more:

- people
- processes
- cows
- information
- equipment
- technology
- suppliers
- customers
- paddocks
- supplement variations
- regulations.

Farm examples of where flow can be improved include:

- grazing plans (see figure 12.3, overleaf)
- flow of people in a shed when milking
- pasture rides
- paddock maintenance flow
- cropping/replanting

343

- calf movement between pens, paddocks and farms
- cow movement between farms
- movement of feed.

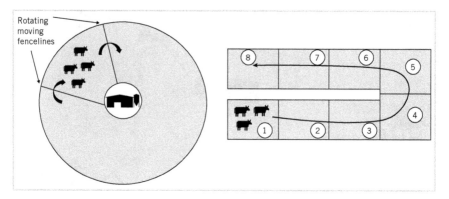

Figure 12.3: examples of creating flow for grazing—imagine how easy this would make your work

Flow of calves during calving is important

Create a better flow on your farm

There are a number of things we can do to make our processes or work flow better:

- *minimise hand-offs* — try to reduce the number of people involved in a particular job so you don't hand off tasks, making the process stop and start

- *continuous flow*—look at your processes and remove batching where possible and any waits and stagnations in the process. Try to make the process achieve one-piece (FIFO) flow

- *straight line flow*—design your process sequence so that it flows in a line to make it easier to manage (see figure 12.4)

- *simultaneous processing*—design your process sequence so that tasks are done in parallel to make the process quicker

- *physical lay-out*—this will probably involve investment but if you can design the process, farm, equipment, facility or building layout with flow in mind, you will be able to establish a very efficient process.

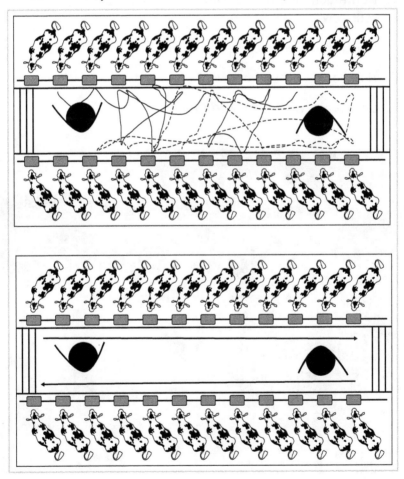

Figure 12.4: spaghetti vs flow in a herringbone shed

Lean Farm team activity

Spaghetti flow

With your team, discuss the following:

1. If you had traced all your steps on the farm yesterday, what would your flow look like?

2. Do you think you could have improved your flow? How?

3. How much time do you think was wasted as a result of your flow?

4. In what areas or processes of the farm could you improve flow?

5. How could flow be improved?

6. Design a new flow for one of these areas.

Flow in a Herringbone Dairy shed can be like spaghetti.

A Lean Farm example

Analysing our calf flow

This year we did a brainstorm with our team of our calf flow across our multiple farms. The easiest way to do this was on butchers paper. Once you draw the process, it is amazing to see what you thought was a pretty good process turn out to be quite messy. As I said, if it looks messy on paper, then it probably is in reality. See figures 12.5 and 12.6 for a comparison on paper.

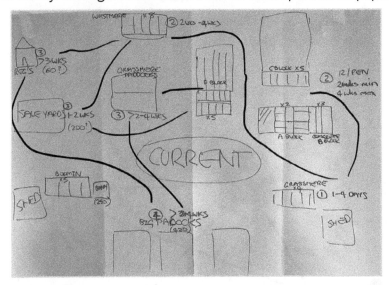

Figure 12.5: our current calf flow at various locations—spaghetti style

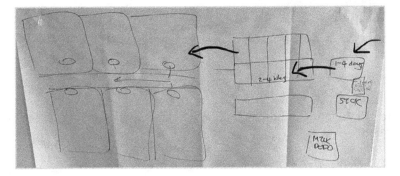

Figure 12.6: our desired future flow—straight line, one location

Information flow

Obviously physical flow is easier to see, draw and analyse. But sometimes it isn't necessarily the physical flow that is the problem, but rather the 'invisible' flow of information (aka communication). If communication between people, systems, suppliers and farms looks like spaghetti, it can cause substantial problems for your farm. Poorly made decisions, or none at all, critical jobs not done, poor timing of jobs and availability are all examples of problems caused by complicated information flow.

The more people involved in your farm business, the more likely information flow is not ideal, as illustrated in figure 12.7.

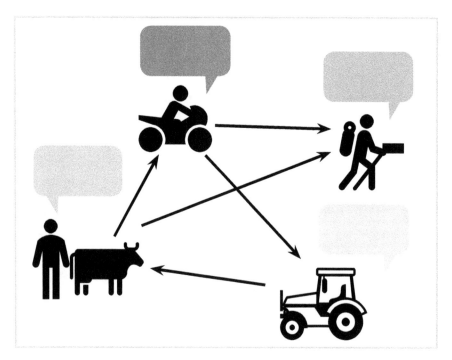

Figure 12.7: farm information flow or communication is often like 'spaghetti'

A Lean Farm example

Too many cooks in the kitchen

Some farmers were explaining to me how a decision is made on their farm where they all worked. I was listening attentively, but after a little while I was confused. We started mapping out the communication process and it became clear how complicated the information flow between various individuals in the business was. Information and decisions went back and forth with multiple people having input. However, instead of all involved parties being in one place, and agreeing on things together at one time, there were multiple simultaneous discussions being made among different groups of people at different times. The result was a very long and inefficient decision-making process, followed by last-minute decisions. While the end result may be successful, it caused a lot of unnecessary stress, frustration and wasted time along the way.

Summary

If you have the opportunity to design new facilities or farms in the future, take note of this chapter. It is a great occasion to think about flow of cows, people, materials, vehicles, information and processes.

In the meantime, think about your existing processes and flows. There are many circumstances and existing processes on the farm where you can introduce JIT systems or at least improve flow of movement to streamline your business.

Sketching out your flow or the flow of a particular process or information can help you visualise just how much spaghetti flow there is. This can also help you start to identify potential improvements in the flow.

Daily movement on farm is often spaghetti flow, with multiple people driving around in multiple farm vehicles

Lean Farm team quiz:
CREATING FLOW ON THE FARM

1. What does JIT stand for?

2. What is the difference between batching and one-piece flow?

3. What does FIFO mean?

4. What does spaghetti flow mean?

5. What is an example of batching on a farm?

Lean Farm action plan:
CREATING FLOW ON THE FARM

These actions are included to provide some simple guidance for your farm. They are aimed at giving you a little bit of motivation and direction if you need it. Of course, the idea is for you to do as much as you can to gain the maximum benefit of creating flow on your farm.

1. Introduce the concept of creating flow to your team.

2. Identify a process that you can change from batch to one-piece (FIFO) flow.

3. Design a solution and implement it if possible.

4. Identify any parts of your farm that suffer from spaghetti flow.

5. Brainstorm how you can introduce smooth flow into this process.

6. What would this new flow look like?

7. If possible, implement the improved flow to the process/area.

8. Sketch your current information flow. Does it look messy? How can you improve it?

9. Take *before* and *after* photos of your old and new flows.

This chapter was a brief introduction to flow. I hope that even though it may be challenging for our farms to create flow due to our infrastructure and land constraints, you can appreciate how flow can significantly simplify your farm processes and improve its productivity.

We have now reached our final chapter on Lean tools—visual planning. This tool is more strategic and will help you put everything you have learned so far into action.

Chapter 13

Visual planning

Planning isn't necessarily a Lean concept, but it plays a significant role in a Lean business. Many farmers I have talked to identify lack of visual planning as a big weakness in their farming business. It's not that they don't want to plan, but they just don't seem to have the time to sit down and do it. It is a vicious cycle, as the more time you spend planning, the more efficient you will be and the more time you will have available. The less time you spend planning, the less efficient you will be and the less time you will have.

In this chapter I am going to briefly introduce a few Lean visual tools that we use on our farm to help us think strategically and plan our core business activities and improvements.

As you can see in figure 13.1 (overleaf), planning is a key part of any successful farm. It helps us to move from our current state to our future or ideal state. Without a plan, how will you know where you want to go and how to get there? Planning, however, needs to be visual to be effective. We might think we do a lot of planning on our farms, but if it is all in our heads, it is very difficult to convey that plan to those who need to execute it.

An effective visual plan must be simple to understand, reviewed regularly, distributed and communicated to everyone, displayed visually and action focused.

Visual planning helps us to:

- identify our priorities
- clarify our goals and vision

- develop a clear roadmap of how to achieve our desired future
- visualise key projects, actions and time frames
- allocate appropriate resources
- monitor progress of our actions
- focus on the right things at the right time
- get things done.

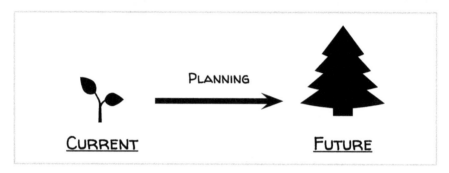

Figure 13.1: planning helps you move from your current to your future state effectively

The Plan–Do–Check–Act (PDCA) cycle

The famous PDCA cycle developed by Dr Deming (also known as the Deming cycle or Shewhart cycle), introduced briefly in chapter 7, is another key component of Lean. The cycle is integral to every action or project that is undertaken at Toyota. Pictured in figure 13.2, PDCA provides a simple, structured approach to working on any action, task or project. The philosophy is that you should always *plan a lot* before going out to do anything.

PDCA is a repetitive, continuous cycle that is used as part of the continuous improvement and standardisation loop. The four steps can be defined as:

1. *Plan*—set a clear plan, including communication and preparation for the task
2. *Do*—test or implement the task, change or project

3. *Check*—monitor the results of the change and verify whether it was successful

4. *Act*—if the change is successful, communicate, share the process and standardise. If the change isn't successful, identify opportunities and plan improvements, repeating the PDCA cycle.

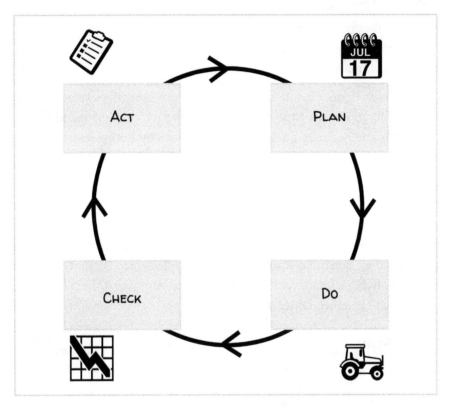

Figure 13.2: the PDCA cycle

The more you plan, the more smoothly the implementation of your actions or projects will go. Therefore your 'doing' time will be quite short with good planning. In a farming environment this is important. If we need to do an important maintenance job on our shed that we weren't able to do during dry-off we can't afford to simply shut down our dairy shed for a week. We need to have a very clear plan of attack so that we can do the work as quickly as possible and the downtime period can be as short as possible to minimise impact on our production.

Most farmers are very good at the 'doing'. Many just want to 'go and get it done'. The planning comes second, if at all. The job still gets done, but not without frustrations such as tools or equipment not being ready or the required people and resources not being available due to last-minute communication mishaps. This then causes delays, rework and often additional cost.

An old colleague of mine at Toyota once drew an excellent model on a napkin over dinner to explain how Lean companies think about planning versus traditional companies. I use it all the time.

The model in figure 13.3 demonstrates that Lean companies spend a long time planning, a short time actually doing the job or implementing the plan, then another long phase of checking and monitoring to make sure that the implementation has generated the expected results, and finally a short period of action.

Traditional companies, on the other hand, spend their time in opposite proportions: on the doing and acting. This is because they don't plan, so everything goes wrong during implementation and it takes much longer, and then they don't check what has been implemented so they spend a long time fixing all the problems. Moral of the story: do it the 'Lean' way and you will be much better off!

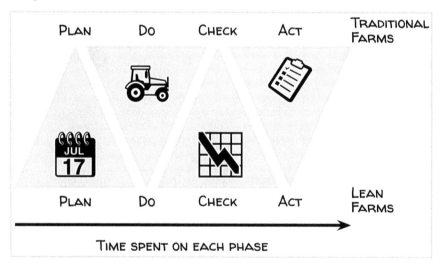

Figure 13.3: the traditional vs Lean approach to planning

A Lean Farm example

If you fail to plan...

Imagine you needed to replace a water tank, and dig in new water piping. However, because there was very little planning done people had other jobs to do and showed up 30 minutes late, the contractor didn't show up at all, a tractor and implements hadn't been moved and were in the way, the digger was at a different farm, the distance wasn't measured accurately and so there wasn't enough length of pipe, and the joiners weren't available until two days later. Furthermore no-one actually knew where the pipes should go and who should do what. This paints a pretty clear picture of how things look without planning, right? And I am sure all of us have experienced similar situations on our farms.

The result: additional time, cost, waiting, waste and frustration

Mat developed a full project plan to manage the construction of our newest dairy shed

Lean Farm team activity

With your team, discuss a recent project or activity you carried out that could have been planned better.

— What went wrong?

— Why was there no or insufficient planning?

— How could planning have made the project or activity easier and better organised?

— What would you do differently next time?

Planning on your farm

One of the most important things you can do for your farm is set an annual and longer term plan. This will help to guide you towards your goals and also achieve the desired key actions. Before planning your annual or shorter term activities, you should be clear about what your farm's longer term objectives are.

Strategic planning

An important Lean system is the 'Hoshin Kanri' process. It is basically a strategic and holistic goal deployment process that is cascaded through the business aligning every person to the business vision, goals and strategy. The first step is to start with a blue sky thinking activity, during which you brainstorm your vision for your farm.

Step 1: blue sky thinking

- What do you want to achieve in the future?
- What are your goals?

- What is your vision?
- What will the farm look like in 10 years' time?

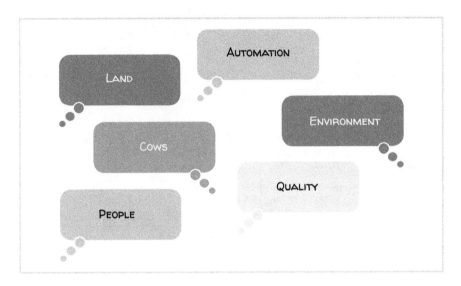

Step 2: five-year strategic plan

Next you should put together a high-level schedule of the key projects or actions you want to achieve over the next five years. This strategic plan should incorporate your blue sky thinking so that it helps you achieve your longer term vision and goals. See the example in figure 13.4.

ACTION	DETAIL	2017	2018	2019	2020	2021
PEOPLE	10 PEOPLE			████████	████████	
	OP MGR		████████			
COWS	HERD 1 500		████			
	HERD 2 650			████		
	NEW HERD					████
LAND	NEW DRY STOCK				████████	
ENVIRONMENT	MONTHLY TEST	████				
	FENCED			████████		
AUTOMATION	UPDATE AFI				████	
QUALITY	SENSORS	████████				
	AUTO WASH		████████			

Figure 13.4: example of a five-year strategic plan

Step 3: annual detailed master schedule

The final step is to develop a detailed annual schedule that focuses on the projects and tasks required over the next year. This captures the next level of detail after the five-year strategic plan. This annual plan should be visualised in your team office and discussed during your team meetings. It should be the guiding tool used by everyone to stay aligned and focused and achieve the annual goals. See the examples in figure 13.5 and 13.6.

cat	Item	Target	Resp	Activity									
				WK 14	WK 15	WK 16	WK 17	WK 18	WK 19	WK 20	WK 21	WK 22	WK 23
DAIRY SHED	VAT system – sensors	0 Milk lost	JH	Investigate options		Agree		Implement		Test			Aud
	Plant set up	100% right first time	BH				Write Procedure		Train All Members		Start		
		0 grades Water reduction	SD										Develo
QUALITY	RFT Culture												

Figure 13.5: example of an annual detailed master schedule

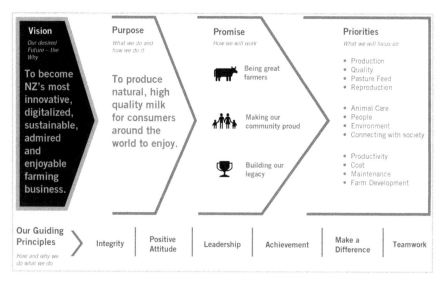

Figure 13.6: Grassmere's five-year strategy

Lean Farm team activity

Blue sky thinking

On your own or with the team, do a blue sky activity. Think about:

— What are your farm or personal goals?

— Where do you want to be in 10 years?

— What do you want your farm and business to look like in the future?

— What is important to you?

Write down some key themes and even targets.

Annual planning

On our farm, we use two main tools to help us plan each year and season: a season plan and a continuous improvement action plan (which is essentially the annual detailed master schedule discussed on the previous page).

Season plan

The first tool is the season plan, which we use to track all key activities required to run the farm over the season as part of normal business operations. The season plan is a simple Excel document that is adjusted as required prior to the start of each season. It is displayed visually on the team visual management boards and the management visual board and referred to weekly to ensure we are on track with all of our core farm activities. The key categories of activities we have on the season plan are People, Feeding, Herd management, Animal health, Young stock, Dairy shed, Pasture (nutrients and irrigation) and Cropping. All of the activities and dates on our season plan (shown in figure 13.7, overleaf) are also scheduled into our electronic shared team calendar (we use Office 365) so that everyone can see what is coming up, wherever they are. Our season plan template is also available at www.leanfarm.nz to help you get started and save you time creating your own.

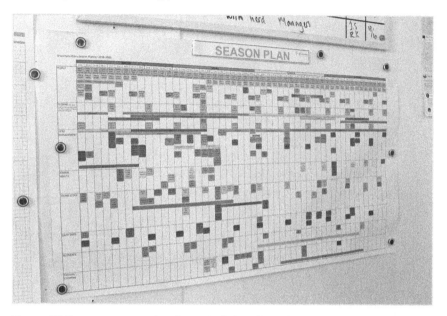

Figure 13.7: our season plan is part of the visual boards in each shed

Continuous improvement action plan (master schedule)

The second key document is our annual master schedule, which we call our continuous improvement action plan (figure 13.8 gives you an example of this). It tracks all the special projects and actions needed to improve the business. These actions are derived from our farm strategy and priorities. (We hold an annual farm strategy day every June with our full team where we discuss the next season's goals and priorities in relation to our longer term vision and strategy.) The plan includes our Lean systems, digitalisation, farm development, people systems, safety and so on. The plan is displayed on the management visual boards and discussed weekly. Items from this continuous improvement action plan are then taken and cascaded down to the team as focus improvement actions for that week or month. These are outside of the normal day-to-day tasks. If you always only focus on the normal day-to-day tasks you will never move forward and make any improvements. Every week we have a one-hour dedicated continuous improvement team session (this

Figure 13.8: an example of our current continuous improvement action plan

EQUIP/FACILITY	#	DATE RAISED	DESCRIPTION/ACTION	WHO	WHEN	JUN	JUL	AUG	SEPT	OCT	NOV	DEC	JAN	FEB	MAR	APR	MAY
PEOPLE	1	1/06	Start using app for all safety recording & contractor management	JS	30/10												
	2	1/06	Shared calendar—add full season plan to it	JH	30/06												
	3	4/06	Staff survey—design and roll out	JH	1/11												
	4	10/06	Review induction process, team feedback and input and simplify	JS	30/12												
	5	10/06	Investigate online interactive induction systems	JH	30/10												
	6	1/06	Cow marking—review and update standard & visualise on board	JS	30/08												
SHED OPERATIONS	7	10/06	Shed preventative maintenance plan—develop & visualise	SC	30/12												
	8	10/06	Develop and trial system for moving cows off platform	MH	30/12												
	9	10/06	Start development of e-learning modules for all plant start up, milking and plant clean process	JH	30/05												
FARM INFRASTRUCTURE/ FACILITIES	10	10/06	Effluent mgmt.—visualise where effluent pipes are and what effluent changing points are	RK	30/05												
	11	10/06	Construct calf loading shed	RK	20/07												
	12	10/06	Apply river run rock to lower traffic races	JS/RK	25/07												
	–	–															

365

is separate from the weekly team meeting), which is used to tackle these improvement actions one at a time in addition to any other available time the team finds in the week.

The key categories we have on our continuous improvement action plan are People, Collaboration, Safety, Management, Shed operations, Farm facilities & infrastructure, Cow management, Environment, and Farm special projects. One of our special projects this season is 'beautification' and as part of that we are planting trees on our farm.

One of our strategic projects: planting trees on our farm

Summary

Planning is critical to help you and your team achieve goals and targets as efficiently and effectively as possible. Of course there are plenty of very successful farms that don't have any planning. But it doesn't mean they don't work extremely long hours and possibly spend a lot of extra money to achieve their success.

Planning will help you achieve your goals in a better, more effective, efficient and organised way. It will ensure that you stay aligned to your vision and strategy. Importantly, planning will provide your team with a better focus and visibility of actions, objectives and progress, creating better commitment and engagement. It is a very powerful management tool for every farm.

Lean Farm team quiz:

VISUAL PLANNING

1. Why is planning important?

2. What does PDCA stand for?

3. What is a blue sky thinking activity?

4. What is the purpose of a schedule?

5. What is meant by visual planning?

Lean Farm action plan:

Visual planning

These actions are included to provide some simple guidance for your farm. They are aimed at giving you a little bit of motivation and direction if you need it. Of course, the idea is for you to do as much as you can to gain the maximum benefit of visual planning on your farm.

1. Do a blue sky thinking activity with your team — what will the farm look like in 10 years?

2. Create a high-level five-year plan for your farm.

3. Brainstorm key projects/improvements/activities that are needed on your farm for the next season.

4. Prepare a master schedule of all your projects and activities for the next year or season.

5. Develop a season plan for your current and next season.

6. Establish a clear review process of your schedules, including frequency of review, who and how.

7. Take *before* and *after* photos of your planning tools.

We have made it to the end of our Lean tools. Of course there are many more Lean tools that I haven't covered in this book. But I think if you can even implement a few of these 10, your farm will have taken a huge leap forward and you will have transformed it!

As you may have recognised throughout the book so far, the success of a lot of these tools relies on your people and bringing them with you so that they are engaged and believe in this. Part III is going to address people, give you a few tools to help you navigate the complexity of human behaviour and mindsets, and use Lean leadership to create the right culture on your farm.

PART III
The culture of a Lean Farm

The technical stuff is easy: all the tools we have talked about can be implemented without too much trouble. What's not so easy is sustaining them and getting your people on board. The people side of things is the most challenging—no matter what business you are in. For farmers the challenge of people is even greater. Most farmers have never been exposed to the numerous tools and skill sets available in the wider corporate business world that help to understand and lead people better.

Farmers aren't HR experts and most farms aren't big enough to warrant a specialist HR department with expertise in 'people and culture'. So it makes it pretty difficult for farmers to have the knowledge and skills they need to help them develop highly engaged, empowered and productive people and teams. For many farmers who are sole operators and don't have teams this may not matter. But for those farmers who have even a handful of employees or want to grow their farming business, the issue of people is one that can't be ignored. Your farm simply won't achieve what you want it to achieve if it relies on you alone because you have a team of people you can't trust, or who don't care or aren't motivated. Farmers who want to lead a team and farm need to develop their people skills so they can truly bring the best out of their team. This will be a win for your team, you and your business. And anyone can learn some simple skills that will make them a better manager and leader.

Furthermore, employees of today have a completely different approach to work and tolerate a lot less of the 'old school' type of management behaviours. This means the way you talk to and manage your people might need to change to ensure you build a positive team culture so that you can achieve all the things we have talked about in this book. While I am by no means an HR or people expert myself, I have spent a lot of time on cultural change and have learned many different techniques and tools to help create and lead high-performing teams. In part III I will share some of these and also explain some of the traits of 'Lean' leaders to help you become a better 'people' person.

Chapter 14

Creating the right culture

The culture of your farm is how it is governed. It entails your business philosophy, vision, values and belief systems. It sets the benchmark of how you operate your business and what is expected of every employee. Creating the right culture on your farm isn't easy—all the Lean tools and technical aspects are easy in comparison. Most businesses can apply the Lean tools to some extent. Where most fail is trying to instil the right culture so that the Lean principles and improvements are sustained.

Having an excellent, well-defined culture on your farm will transform it. It will have many benefits including:

- having a highly motivated, engaged and committed team
- improving productivity and performance
- having an enjoyable working environment
- achieving a better work–life balance
- being a safe workplace
- improved employee retention
- recruitment of people who fit your culture.

All of this will ultimately allow your farming business to achieve its goals and targets successfully. This is where we are trying to spend most of our time and energy on our farm. We want to be a leader and the most admired farm in the future. From my experience if you have the right culture in your business, everything else just follows.

As I've mentioned, introducing Lean to your farm is a cultural change. You might be successful in introducing a few of the tools, and you will see some benefits, however unless you get everyone in your team to change their mindsets it is unlikely that you will create a truly continuously improving farm.

Toyota's culture

I have worked with dozens of businesses from large global corporates to small family farms. Yet I have never forgotten the strong culture that existed at Toyota. Out of all the companies I have worked with and for, Toyota had by far the strongest, most defined and real culture. You lived and breathed it every day—it became the DNA of every single person. Toyota has a very powerful culture and I can still feel it today.

It is Toyota's culture (known inside Toyota as 'the Toyota Way') that makes the Lean systems (the Toyota Production System) work and remain sustainable. Without this strong culture, many of the Toyota production systems and tools would not be so effective. They are reliant on their people's dedication, passion and genuine pursuit for perfection and continuous improvement.

Toyota's culture is based on two pillars of continuous improvement and teamwork (see figure 14.1). Each of these pillars has several elements. You can find out about these pillars and cultural elements on the Toyota Europe website.

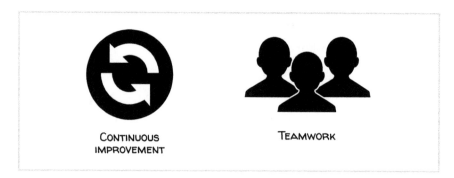

CONTINUOUS IMPROVEMENT   TEAMWORK

Figure 14.1: the two pillars of the Toyota Way

Continuous improvement is about creating an environment where people are encouraged to challenge the status quo and seek improvements in everything they do. To be able to do this you need a culture that promotes respecting each other, listening to others, and that supports ownership and empowerment. This is the essence of the teamwork pillar. Without teamwork and basic respect for each other, people will not be confident or feel supported to raise problems and ideas and to challenge the norm. Therefore continuous improvement would not be possible.

Your farm's values

In order to create a culture, you first need to define the culture that you want for your farm. Your values and vision will generally guide this. Your farm values should express what is important to you and the guiding principles under which you want to operate your business.

On our farm, teamwork is a core value

Lean Farm team activity

Defining your farm's values

On your own or with your team, discuss the following:

1. Do you have farm values? What are they?

2. If not, what are three values you would like to have? What kind of culture would you like to have?

3. Do your values mean something to you? Do you believe in them?

4. Will they achieve the culture you want?

We did a team brainstorming session a couple of years ago to develop our farm values. We did it with the team so that we could establish values that not only were important to us, but also that people could relate to.

Our values are:

- *Integrity*—we are reliable; we know right over wrong; we are ethical in our dealings at all times; we don't cut corners.
- *Positive attitude*—we are friendly; we are 'can-do'; we are positive and not boring.
- *Leadership*—we inspire others; we lead by example; we take responsibility and ownership.
- *Achievement*—we set goals and are committed to achieving them; we follow through and earn the rewards.
- *Make a difference*—we make a difference in our community and our environment, we think outside the box, we are creative and innovative.
- *Teamwork*—underpins all our values and is our strength in good times and bad.

Once you define your values, it will be much easier for you to set expectations around the behaviours you expect within your team and business. As behaviours are hard to change, setting expectations up front —for example, during your recruitment process — will enable you to attract and recruit people who can associate with your values from the start. Chapter 15 discusses behavioural patterns of people to help you understand your team and their behaviours better.

Chapter 15
People behaviours

To change or create a culture, you need to change people's mindsets and behaviours. To do this it is good to have a basic understanding of how people behave and why. I most certainly am not a change specialist or a psychologist. As a Lean consultant, however, I spend a lot of my time leading change and trying to change cultures and I have learned some things along the way.

CULTURES DON'T CHANGE — INDIVIDUALS CHANGE.

One interesting concept I came across is that you can't change a culture unless you change a lot of individuals. Fortunately, most farms only have a few employees, not hundreds or thousands. This makes it much easier to focus on each individual and work with them to slowly change behaviours. Better yet, if you have a well-defined culture, you can explain this during your recruitment process, and therefore recruit people who will fit in with this culture and your values.

Many Lean companies know that technical skills are easy to teach. However, behaviours and values are difficult to change. Therefore many companies recruit more on behaviours and value systems than technical skills. This doesn't mean you should recruit someone without a practical bone in them and expect them to repair tractors straight away. You need to use common sense, of course. We have employed many team members with no hands-on experience of farming. Those with the right values and good work ethic who were prepared to learn and give things a go have been excellent members of our team, and have developed their technical skills quickly.

Understanding people

There are a few models I have learned about over the years that have been very helpful to better understand how people work. Change is a difficult process for many. Introducing Lean thinking, a new culture or any other change into your farm can be challenging for people. People will react differently to any type of change. For the change to be successful you need everyone on board, engaged and wanting to be involved and drive the change. If people aren't on board, the change will be very difficult to implement. To bring people along with you, it is good to understand and be able to recognise the various reactions they have to change. This will enable you to manage the change more appropriately and successfully.

Joey's wife Sara giving our team a boost during calving

Types of behaviour

There are generally four modes of behaviour or reactions that people can express when confronted with change. Most of us, if not all, have experienced being in every one of these modes at some point in our lives. This is because we react differently to different changes. Sometimes we agree with the change and other times we don't want it at all. How people react to a particular change often relates to their past experiences, which we may not know anything about. This creates psychological filters (we will talk about these next). So what are these types of behaviour? The four modes are Critic, Victim, Bystander and Navigator.

Critic

Negative about change and challenges it. Thinks there is no need and has a 'hasn't worked before' attitude.

Victim

Resists change, wants things to stay the same, tries to go back to old ways of doing things, asks 'Why me?'

Bystander

Tries to ignore it and hope it goes away. Doesn't want to be involved. Waits for others to lead.

Navigator

Engaged and wants to lead change. Optimistic and sees it as a good opportunity for improvement.

* * *

By identifying which mode of behaviour a person is in early on, you can approach the situation the right way and more effectively. Having the right discussions and addressing any issues immediately can help the person understand and manage the change better, and ideally transition to navigator mode.

Lean Farm team activity

Types of behaviour

As a team, discuss a recent change you have had.

— What was your reaction?

— Which mode were you in?

— Think about past changes: have you shown other modes of behaviour too?

— How did you overcome your initial reaction?

— Did you accept the change?

Psychological filters

Every person has a history and a databank of past experiences, knowledge, encounters and situations that make them perceive things their own way. Our past baggage makes us interpret information, communication and change according to these past experiences. We essentially have internal psychological filters built up from our previous experiences and our initial reactions are a reflection of these filters (see figure 15.1).

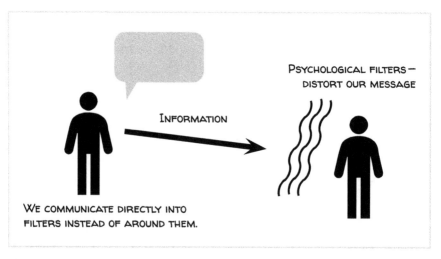

Figure 15.1: we communicate directly into filters instead of around them

Therefore, if we are trying to communicate with people, our messages are heard and then translated or manipulated in that person's mind based on their internal psychological filters. As a result, our messages could be blocked out or distorted somehow.

This is important to understand because sometimes we think what we're saying makes perfect sense and it is completely logical, and we just can't understand why others don't agree with us and want to jump on board.

If we appreciate this notion of psychological filters, we can better understand why people don't have the reactions we expect when we communicate a message. If we can try to understand what is going on in someone's head, this can help us communicate in different ways or formulate our message in a way that gets past the psychological filters to ensure that the right message is delivered.

Right. So now you have a better understanding of how our brains work and why we react to things the way we do. The final chapter of this book is dedicated to you, the leader. You will be pivotal in making Lean happen and stick. Therefore you also will need to be equipped with some new skills to help you become a different kind of leader—a Lean leader.

Chapter 16

Lean leadership

Nothing can change unless the right leadership is in place. The information in this chapter is probably the most difficult for farmers as it requires quite different thinking and a change in the way they manage and work with their people. It requires farm owners, sharemilkers, managers and others to move from the standard 'management' approach to 'leadership'. If we use the right leadership approach on our farm, we will be much more successful in creating an engaged, high-performing team and farm.

Leadership behaviours

A Lean leader demonstrates very specific behaviours and styles of leadership. One of the key Lean leadership traits is that the leader's fundamental role is to support the entire team and business, enabling the team to do what they need to do. We essentially spin the traditional hierarchy upside down so that the leader is at the bottom supporting everyone else. We shift from a 'command and control' style of leadership to that of 'empowerment and support' (see figure 16.1, overleaf).

Where are you spending your time?

If you want to lead your team, you also need to change where you spend your time. This can be quite a challenge for most farmers, particularly owners. If you are spending all your day working in the business milking, fencing, moving stock and solving everyone's problems, you aren't being a leader.

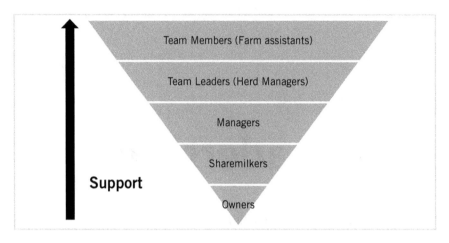

Figure 16.1: a Lean leader's role is to support their team

As a leader you need to spend the majority of your time firstly supporting your team by coaching them and developing their knowledge and skills, and secondly on more strategic activities. Where do you currently spend your time in a day?

Traditional	Lean
— Micromanagement	— Leading
— Day-to-day tasks	— Supporting
— Can't step out of business	— Empowering team
— Firefighting	— Planning
— Solving everyone's problems	— Coaching
	— Working on the business
	— Developing people

Traits of Traditional vs Lean leaders

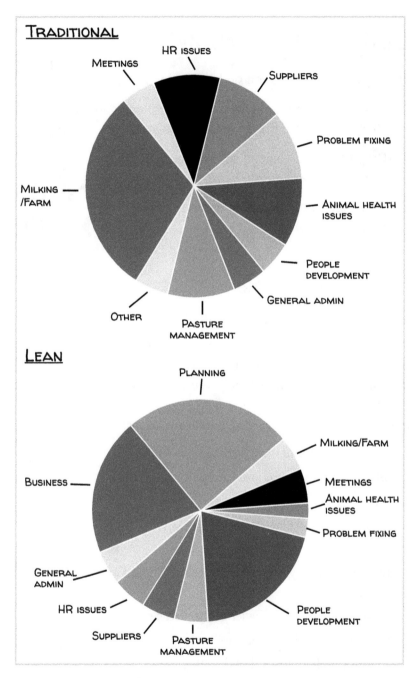

Figure 16.2: how traditional vs Lean leaders spend their time

Lean Farm team activity

Track all the activities you do by minutes for one day.

1. Break down your activities into several categories like those shown in figure 16.2 on previous page.

2. Graph them so you can see where you spend your day.

3. How much time do you spend on leadership tasks?

Empowering people

As hard as it may seem, farmers need to get better at empowering people. If you are constantly feeling like you have to do everything, and nothing gets done without you, it's time to shift to Lean leadership. All the Lean tools that we have talked about will help you to stop working in the business trying to do everything and be involved in everything, and instead start to empower your team. Empowering your team means:

- giving them autonomy to take actions
- listening to their ideas and feedback
- involving them in key decisions
- letting them make decisions
- asking them for their input
- recognising that your team have valuable input and can contribute to the business
- encouraging them to challenge the status quo
- giving them further responsibilities
- supporting them instead of directing them

- developing their knowledge and capability
- making them the masters of their jobs
- celebrating their achievements
- encouraging their participation
- helping them think for themselves through coaching.

The Lean tools detailed in part II provide your team with clear structure and understanding to know what to do, to make the right decisions and to take the right action. The tools will also enable you to see visually what is happening without being involved and give you confidence in the team (see figure 16.3, overleaf). By empowering people you will have more time to:

- work on the business, not in the business
- focus on strategic projects
- grow your business
- develop a high-performing team
- develop yourself
- have a better work–life balance
- take a holiday!

Ultimately, this will result in a more productive, successful farm.

Lean Farm team activity

Empowering your team

1. What kind of leadership style do you have?

2. Do you empower your team?

3. How do you empower your team?

4. Does your team feel empowered?

5. What can you improve?

Figure 16.3: Lean tools will help you to empower your team

Moving to coaching

Traditionally, farmers are very directive in their management style, usually telling people what to do. A key way to develop your people so that they are empowered is to use a 'coaching' style instead. I am not talking about the 'sport' coaching style but the 'bringing out the best in your people' coaching style, as depicted in figure 16.4.

In sport, when you say 'coach' it means someone (the coach) gets out on the field (or in the change rooms) at half time and gives pretty clear instructions (sometimes in the form of shouting) about what needs to change, how to do it and who needs to do what.

Figure 16.4: the two types of 'coaching'

Instead, we want to develop our people to think for themselves. Rather than telling them what to do all the time, we want to ask them for their opinions and suggestions.

To coach your team, you need to get into the habit of asking open-ended questions, to encourage each person to think and work problems out in their own head. We all know that when we have been left to do something on our own and had to work it out ourselves, we learnt the task much more quickly, developed confidence and became masters at it. When someone just directs you what to do and how, you simply go and do what you are told with minimal understanding and thinking. By asking closed-ended questions that require only a 'yes' or 'no' answer, you won't teach your team to think, develop them or empower them.

Moving to a coaching style of leadership will result in big wins for your farm, including:

- a team of excellent problem solvers
- a motivated team
- an empowered team
- people thinking for themselves
- improved performance
- better decision making
- challenging the status quo to improve.

Lean Farm team activity

Coaching

Ask your team members to find a partner to work with. One person is the coach and the other has a problem to fix.

1. Identify a current problem that you each have.

2. The coach should spend five minutes asking *only* open-ended questions to try and 'pull' a solution from the other person.

3. Swap roles, with the other person now acting as the coach. Repeat step 2.

4. How difficult was that? Did you try to *tell* the person what they should do?

Motivating people

In a 'command and control' style of work environment, it is common to see an unmotivated team that is not performing. On the other hand, in a Lean leadership 'coaching' style of work environment, people are empowered, engaged and motivated. These businesses perform above minimum expectation. If people are performing below what I call the red line (minimum expectation of performance), you will need to use a combination of coaching and constructive feedback. This will help you raise their performance to at least the minimum expectation.

If, however, a team member is already performing above the minimum expectation, it is not suitable to use 'command and control' or constructive feedback to further improve their performance. This will just anger the team member and make them feel unvalued. In this situation you should only use coaching to motivate that team member to achieve their absolute potential. This will make the person feel valued, empowered and motivated. And your farm will achieve exceptional culture and performance. Figure 16.5 expresses this diagrammatically.

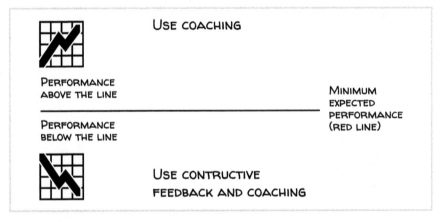

USE COACHING

PERFORMANCE
ABOVE THE LINE

MINIMUM
EXPECTED
PERFORMANCE
(RED LINE)

PERFORMANCE
BELOW THE LINE

USE CONTRUCTIVE
FEEDBACK AND COACHING

Figure 16.5: improving the performance of your people

Summary

Adapting your leadership style to become a Lean leader will be the most challenging aspect of this book but also one of the most important and rewarding. Any change and improvement starts with leadership, and it is leadership that will determine its success or failure. As hard as it is to sometimes acknowledge, when failures occur or you don't achieve the expected results, you need to look at yourself first as a leader before blaming others. By taking on board some of the techniques that we have discussed in this section and starting to apply them in your management approach, you will become a better leader and you will be rewarded with a more engaged team and a better farming business.

Lean Farm team quiz:

CREATING THE RIGHT CULTURE

1. What are the four modes of behaviour?

2. What does coaching mean?

3. What is the difference between 'telling' and 'asking'?

4. How should you manage performance if it is above the red line?

5. When should you use constructive feedback?

6. What is the role of a leader?

Lean Farm action plan:
CREATING THE RIGHT CULTURE

These actions are included to provide some simple guidance for your farm. They are aimed at giving you a little bit of motivation and direction if you need it. Of course, the idea is for you to do as much as you can to gain the maximum benefit out of creating the right culture on your farm.

1. Introduce 'creating the right culture' to your team.

2. Decide what you want the culture of your farm to look like.

3. Agree on your farm values and display these visually.

4. Include your values and culture in your recruitment process.

5. During your next change, identify the four modes of behaviour among your team and discuss.

6. Start to move towards Lean leadership—prioritise your activities so that you spend more time working on the business.

7. Establish an action plan for starting to empower your team more. Ask your team for their feedback.

8. Instead of giving answers, start to use only open-ended questions with your team (coaching).

You have made it to the end. Well done! I hope that this final chapter has given you some more confidence and skills to address your challenging people issues. If you do at least some of the things in this chapter, you will be in a much better position to implement the 10 Lean tools successfully. I am confident that what you have learned in this book will be useful to you and your team and will help you take your farm on its improvement journey. I wish you all the best in transforming your farm to a Lean Dairy Farm. If you know of anyone who could benefit from this book, please share it with them.

About the authors

Jana is a Lean consultant with more than 17 years' experience working for Toyota and in Lean consulting. She is the founder of consulting company Improve8. Jana has helped hundreds of companies around the world use Lean thinking to help them improve their businesses. Since 2013 she has also been involved in supporting her husband's 1000-head dairy farm in New Zealand, introducing Lean thinking to the business. She developed and rolled out the successful LeanFarm training program across New Zealand in 2017. To find out more about Improve8 or Lean Farming, or to purchase any of the templates discussed in this book, visit www.improve8.com or www.leanfarm.nz.

Mat returned to his 1000 head family dairy farm in New Zealand in 2013, after spending more than 10 years living and working overseas in international policy. He is a fourth generation farmer and has overseen the continued expansion of the farm. In 2017 he was made a Nuffield scholar and traveled to 14 countries researching innovation in agriculture and food. Mat has embraced Lean thinking on the farm and has seen the benefits Lean can bring to the team and business first-hand.

Index